Edith Wharton

The Age of Innocence

純真年代

商務印書館

This Chinese edition of *The Age of Innocence*
has been published with the written permission of
Black Cat Publishing.

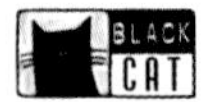

Name of Book: The Age of Innocence
Editor: Michela Bruzzo
Design and art direction: Nadia Maestri
Computer graphics: Simona Corniola
Picture research: Laura Lagomarsino
Edition: ©2008 Black Cat Publishing,
an imprint of Cideb Editrice, Genoa, Canterbury

系 列 名：Black Cat 優質英語階梯閱讀 · Level 5
書　　名：純真年代
責任編輯：畢　琦
封面設計：張　毅
出　　版：商務印書館（香港）有限公司
香港筲箕灣耀興道 3 號東滙廣場 8 樓
http://www.commercialpress.com.hk
發　　行：香港聯合書刊物流有限公司
香港新界大埔汀麗路 36 號中華商務印刷大廈 3 字樓
印　　刷：中華商務彩色印刷有限公司
香港新界大埔汀麗路 36 號中華商務印刷大廈
版　　次：2009 年 3 月第 1 版第 1 次印刷
© 2009 商務印書館（香港）有限公司
ISBN 978 962 07 1856 4
Printed in Hong Kong

出版說明

本館一向倡導優質閱讀，近年來連續推出了以"Q"為標識的"Quality English Learning 優質英語學習"系列，其中《讀名著學英語》叢書，更是香港書展入選好書，讀者反響令人鼓舞。推動社會閱讀風氣，推動英語經典閱讀，藉閱讀拓廣世界視野，提高英語水平，已經成為一種潮流。

然良好閱讀習慣的養成非一日之功，大多數初中級程度的讀者，常視直接閱讀厚重的原著為畏途。如何給年輕的讀者提供切實的指引和幫助，如何既提供優質的學習素材，又提供名師的教學方法，是當下社會關注的重要問題。針對這種情況，本館特別延請香港名校名師，根據多年豐富的教學經驗，精選海外適合初中級英語程度讀者的優質經典讀物，有系統地出版了這套叢書，名為《Black Cat 優質英語階梯閱讀》。

《Black Cat 優質英語階梯閱讀》體現了香港名校名師堅持經典學習的教學理念，以及多年行之有效的學習方法。既有經過改寫和縮寫的經典名著，又有富創意的現代作品；既有精心設計的聽、説、讀、寫綜合練習，又有豐富的歷史文化知識；既有彩色插圖、繪圖和照片，又有英美專業演員朗讀作品的 CD。適合口味不同的讀者享受閱讀之樂，欣賞經典之美。

《Black Cat 優質英語階梯閱讀》由淺入深，逐階提升，好像參與一個尋寶遊戲，入門並不難，但要真正尋得寶藏，需要投入，更需要堅持。只有置身其中的人，才能體味純正英語的魅力，領略得到真實的快樂。當英語閱讀成為自己生活的一部分，英語水平的提高自然水到渠成。

商務印書館（香港）有限公司
編輯部

使用說明

1 應該怎樣選書？

按閱讀興趣選書

《Black Cat優質英語階梯閱讀》精選世界經典作品，也包括富於創意的現代作品；既有膾炙人口的小說、戲劇，又有非小說類的文化知識讀物，品種豐富，內容多樣，適合口味不同的讀者挑選自己感興趣的書，享受閱讀的樂趣。

按英語程度選書

《Black Cat 優質英語階梯閱讀》現設Level 1 至 Level 6，由淺入深，涵蓋初、中級英語程度。讀物分級採用了國際上通用的劃分標準，主要以詞彙（vocabulary）和結構（structures）劃分。

Level 1 至 Level 3 出現的詞彙較淺顯，相對深的核心詞彙均配上中文解釋，節省讀者查找詞典的時間，以專心理解正文內容。在註釋的幫助下，讀者若能流暢地閱讀正文內容，就不用擔心這本書程度過深。

Level 1 至 Level 3出現的動詞時態形式和句子結構比較簡單。動詞時態形式以簡單現在式（present simple）、現在進行式（present continuous）、簡單過去式（past simple）為主，句子結構大部分是簡單句（simple sentences）。此外，還包括比較級和最高級（comparative and superlative forms）、可數和不可數名詞（countable and uncountable nouns）以及冠詞（articles）等語法知識點。

Level 4至 Level 6 出現的動詞時態形式，以現在完成式（present perfect）、現在完成進行式（present perfect continuous）、過去完成進行式（past perfect continuous）為主，句子結構大部分是複合句（compound sentences）、條件從句（1st and 2nd conditional sentences）等。此外，還包括情態動詞（modal verbs）、被動式（passive forms）、動名詞

（gerunds）、短語動詞（phrasal verbs）等語法知識點。

根據上述的語法範圍，讀者可按自己實際的英語水平，如詞彙量、語法知識、理解能力、閱讀能力等自主選擇，不再受制於學校年級劃分或學歷高低的約束，完全根據個人需要選擇合適的讀物。

2 怎樣提高閱讀效果？

閱讀的方法主要有兩種：一是泛讀，二是精讀。兩者各有功能，適當地結合使用，相輔相成，有事半功倍之效。

泛讀，指閱讀大量適合自己程度（可稍淺，但不能過深）、不同內容、風格、體裁的讀物，但求明白內容大意，不用花費太多時間鑽研細節，主要作用是多接觸英語，減輕對它的生疏感，鞏固以前所學過的英語，讓腦子在潛意識中吸收詞彙用法、語法結構等。

精讀，指小心認真地閱讀內容精彩、組織有條理、遣詞造句又正確的作品，着重點在於理解"準確"及"深入"，欣賞其精彩獨到之處。精讀時，可充分利用書中精心設計的練習，學習掌握有用的英語詞彙和語法知識。精讀後，可再花十分鐘朗讀其中一小段有趣的文字，邊唸邊細心領會文字的結構和意思。

《Black Cat 優質英語階梯閱讀》中的作品均值得精讀，如時間有限，不妨嘗試每兩個星期泛讀一本，輔以每星期挑選書中一章精彩的文字精讀。要學好英語，持之以恆地泛讀和精讀英文是最有效的方法。

3 本系列的練習與測試有何功能？

《Black Cat 優質英語階梯閱讀》特別注重練習的設計，為讀者考慮周到，切合實用需求，學習功能強。每章後均配有訓練聽、說、讀、寫四項技能的練習，分量、難度恰到好處。

聽力練習分兩類，一是重聽故事回答問題，二是聆聽主角對話、書信朗讀、或模擬記者訪問後寫出答案，旨在以生活化的練習形式逐步提高聽力。每本書均配有CD提供作品朗讀，朗讀者都是專業演員，英國作品由英國演員錄音，美國作品由美國演員錄音，務求增加聆聽的真實感和感染力。多聆聽英式和美式英語兩種發音，可讓讀者熟悉二者的差異，逐漸培養分辨英美發音的能力，提高聆聽理解的準確度。此外，模仿錄音朗讀故事或模仿主人翁在戲劇中的對白，都是訓練口語能力的好方法。

閱讀理解練習形式多樣化，有縱橫字謎、配對、填空、字句重組等等，注重訓練讀者的理解、推敲和聯想等多種閱讀技能。

寫作練習尤具新意，教讀者使用網式圖示（spidergrams）記錄重點，採用問答、書信、電報、記者採訪等多樣化形式，鼓勵讀者動手寫作。

書後更設有升級測試（Exit Test）及答案，供讀者檢查學習效果。充分利用書中的練習和測試，可全面提升聽、説、讀、寫四項技能。

4 本系列還能提供甚麼幫助？

《Black Cat 優質英語階梯閱讀》提倡豐富多元的現代閱讀，巧用書中提供的資訊，有助於提升英語理解力，擴闊視野。

每本書都設有專章介紹相關的歷史文化知識，經典名著更附有作者生平、社會背景等資訊。書內富有表現力的彩色插圖、繪圖和照片，使閱讀充滿趣味，部分加上如何解讀古典名畫的指導，增長見識。有的書還提供一些與主題相關的網址，比如關於不同國家的節慶源流的網址，讓讀者多利用網上資源增進知識。

Contents

END These symbols indicate the beginning and end of the passages linked to the listening activities. 聽力練習所涉段落開始和結束的標記

第 1, 3, 4, 6, 7, 8, 9, 10, 11 章錄音載於隨書附贈光碟；第 2, 5 章和《離婚簡史》的錄音可從以下網站下載：www.blackcat-cideb.com。

Edith Wharton in 1905, the year her first successful novel, ***The House of Mirth***, was published.

The Life of Edith Wharton

Edith Wharton (1862-1937) was born Edith Newbold Jones. Her family was very rich and well known in New York high society. In 1885, at the age of twenty-three, Edith married Edward Wharton, a banker, whom she divorced in 1913. Her first book, *The Decoration* [1] *of Houses* (1897), was a non-fiction volume co-written with Ogden Codman Jr. In it she criticized the taste and the social snobbery [2] of her parents' generation of the New York high society. These were later to become major themes in her fiction. She published a collection of stories entitled *The Greater Inclination* in 1899 and a short novel – *The Touchstone* – in 1900. *The Valley of Decision* (1902), her first full-length novel, was set in Italy in the eighteenth century. *The House of Mirth* – her first popular success – followed in 1905.

In 1907 she moved to Paris and joined a group of American expatriate [3] artists. *Ethan Frome* appeared in 1911 and was a great success. She

1. **Decoration** : a thing that makes sth look more attractive.
2. **snobbery** : the attitudes and behaviour of sb who believes he is better than other people.
3. **expatriate** : living in a country that is not one's own.

returned to her usual concerns, criticizing the hypocrisies [1] of New York society with *The Reef* (1912) and *The Custom of the Country* (1913). She did a great deal of work to help refugees in France during World War I, for which she received the Legion of Honor in 1916 and the Order of Leopold in 1919. In all, she wrote seventeen novels, eleven volumes of short stories, an autobiography, a volume of poetry, a book of travel writing, a book on Italian villas, two volumes of essays on Italy, a collection of articles on her war experiences in France, and a book on the writing of fiction. At the time of her death, at the age of seventy-five, she was working on a new novel, *The Buccaneers*, set in Saratoga, New York, in the 1860s.

The Age of Innocence appeared in 1920 and was awarded the Pulitzer Prize, which had never before been given to a woman. Wharton's writing is characterized by wit, satire, irony, and a sense of the pain inflicted [2] by social conventions. Like Henry James, she focuses on the life of the emotions in a society that forbids the free expression of passion. *The Age of Innocence* explores these central concerns in great depth and with remarkable delicacy [3] and subtlety [4].

1 Comprehension check

Say whether the statements are true (T) or false (F), and correct the false ones.

		T	F
1	Edith Wharton's family had very little money.	☐	☐
2	Her first book was a novel about the New York high society.	☐	☐
3	Her books were never very popular with the general public.	☐	☐
4	She spent nearly half of her life away from the United States.	☐	☐
5	She wrote a wide variety of books.	☐	☐

1. **hypocrisies** : behaviour in which sb pretends to have moral standards or beliefs that they do not actually have.
2. **inflicted** : make sb/sth suffer sth unpleasant.
3. **delicacy** : the quality of being done carefully and gently.
4. **subtlety** : sth that is important but difficult to notice.

The Characters

From left to right: **Sillerton Jackson, Larry Lefferts, Mrs van der Luyden, Julius Beaufort, Newland Archer, Ellen Olenska, May Welland , Mrs Archer, Mrs Welland.**

Before you read

1 Discussion

You are going to read a story about the rich ruling families of New York in the 1870s. This was a rather small group of families and everybody knew everybody else and everybody gossiped about everybody else. A school is also a closed social group, and gossip is one of its important social mechanisms. With your friends discuss the following questions. Then write down your ideas in a report.

What do you notice about your fellow students?

- who they are dating
- how they act in class
- what jobs their parents have
- what they wear
- what grades they get
- who their friends are

- Do you gossip about these things?
- Do others in your school gossip about these things?
- Do you think gossip is always bad?

2 Listening

You will hear about an evening at the New York Academy of Music; the high society of the city have come to listen to the opera *Faust* and, most of all, observe each other. For questions 1-6, complete the sentences.

Mrs Mingott does not go the opera because she **(1)** ....

May is Mrs Welland's **(2)** ....

Newland thinks that May does not really **(3)** .... the opera that she is watching.

May and Newland have not announced that **(4)** ....

On stage, the singer Madame Nilsson is singing a **(5)** ....

Larry Lefferts knows everything about **(6)** ....

CHAPTER **ONE**

At the Opera

High society [1] in New York in the early 1870s was a very small world. Everyone knew everyone else's business. They all went out in the evenings, dressed in their finest clothes, to attend the theater, the ballet, or the opera; to visit people and to be seen. They gossiped about upcoming marriages and recent scandals. The ladies approved or disapproved of one another's dresses and hairstyles. Hundreds of pairs of eyes watched out for something irregular, something scandalous or ridiculous, and hundreds of tongues were ready to talk about it.

One January evening, Newland Archer went to the opera. The famous soprano Christine Nilsson was singing in *Faust* [2] at the New York Academy of Music. Everyone was there. [3] As Madame

1. **High society** : (here) the rich old families of New York.
2. ***Faust*** : 1859, opera by Gounod about Faust, an old man who sells his soul to the Devil for joy and power.
3. **Everyone was there** : (here) "Everyone in high society was there".

Nilsson was singing a passionate love song, Newland looked over at Mrs Manson Mingott's box [1]. Mrs Mingott herself was far too old and fat to go to the opera, but her family used the box. Tonight her daughter Augusta Welland was there beside her sister-in-law, Mrs Lovell Mingott. Behind them sat a young woman in a white dress. This was Mrs Welland's daughter, May. She was staring at the love scene on the stage. Her eyes were bright, and she was blushing [2], as the blonde soprano [3] sang out "M'ama!" [4] triumphantly.

"The darling!" thought Newland with pride and satisfaction. "She doesn't even know what it's all about. When we're married, we'll read *Faust* together by the Italian lakes." That afternoon, he and May had told each other their feelings. They were now engaged to be married, although they hadn't yet made a formal announcement. He was glad that May was innocent, but once they were married he'd educate her. When she was his wife, he'd teach her to be charming [5] and sophisticated, like the married woman who had fascinated him for two years. He wanted May to have all that woman's charm but none of her weaknesses. Newland looked back at the stage, where Madame Nilsson was reaching the climax of her love song.

Larry Lefferts and Sillerton Jackson were standing next to Newland. Mr Jackson was an old society gossip: he knew the secrets and scandals of all of New York society for the past fifty years. Lefferts was an elegant young gentleman, an expert on

1. **box** : (here) a balcony overlooking the stage in an opera house. Boxes are expensive and only the rich can afford them.
2. **blushing** : her face was red with excitement or embarrassment.
3. **soprano** : a singer whose singing voice is the highest.
4. **"M'ama!"** : *(Italian)* he loves me!
5. **charming** : fascinating; having attractive manners.

what was appropriate and inappropriate behavior in New York high society. "How's the law, Archer?" Lefferts asked a little ironically. Newland was a lawyer in a distinguished New York law firm, but everyone knew that he didn't care much about his work.

"It's all right," said Newland. "A little dull [1], but a gentleman must do something, so I go to Mr Letterblair's office every morning."

Suddenly, Lefferts, who had been looking at the people in the boxes opposite, said, "My God!"

Newland saw that he was staring at Mrs Mingott's box. Another lady had just entered it – a slim young woman wearing a band of diamonds in her dark hair and a very elegant Empire-style dress [2]. Everyone in the opera house was looking at that dress. "Augusta Welland shouldn't have brought her here," said Lefferts.

Newland said nothing but in his heart he agreed. He was a generous young man, and he was glad that May and her family were kind to her unfortunate cousin Countess Olenska. But being kind to her at home was one thing: bringing her to the opera was another thing entirely. Mrs Welland shouldn't have done it.

"What happened to the Countess?" asked a young man close by. "Everyone says that she's 'unfortunate', but I've never heard her story."

"She left her husband," replied Lefferts.

"I heard her husband was horrible," said the young man, who obviously wanted to defend the lady.

1. **dull** : boring.
2. **a very elegant Empire-style dress** : a dress in the style used by Josephine, the wife of Napoleon Bonaparte, with a very high waist and neckline.

"Yes, he was," Lefferts agreed.

The young man looked satisfied, but then Lefferts added, "She ran away with his secretary."

"Oh dear!" said the young man.

"It didn't last long, though," said Mr Jackson. "Last month she was living alone in Venice. Lovell Mingott went there and brought her home. That's fine — a family should take care of its unfortunate members — but bringing her to the opera is a mistake."

"Especially with Miss Welland,"[1] said Lefferts.

Newland suddenly wanted to go to Mrs Mingott's box, to show the world that he was engaged to May, and to protect her from any difficulties she might have as a result of her cousin's scandalous reputation. He hurried through the red corridors to the other side of the opera house. When he entered the box, his eyes met May's, and he saw that she instantly understood his motive.

"Do you know my niece, Countess Olenska?" asked Augusta Welland.

Newland had not seen the Countess since she was little Ellen Mingott — a lively, pretty child of nine. Ellen's parents had liked traveling. When Ellen was little, they took her all over Europe. They died when she was nine, and her aunt Mrs Medora Manson took care of her after that. Mrs Manson was also a traveler. Occasionally[2] she came back to New York with a new husband. Shortly after the death of Ellen's parents, Mrs Manson brought her niece to New York. New York society was shocked to see that

1. **Especially with Miss Welland** : the young man thinks that Miss Welland — a respectable young unmarried woman — should not be seen in public with Countess Olenska.
2. **Occasionally** : sometimes but not often.

the little girl wasn't wearing black, even though her parents had died recently. Instead she wore bright red silk and amber [1] beads.

For a few months, Newland had seen her often in the houses of her aunts or his, but then Mrs Medora Manson had taken her back to Europe. Nothing was heard of them for ten years, then there was news: Ellen had married a very rich Polish nobleman she had met at a ball in Paris. Apparently the Count had beautiful houses in Paris, Nice, and Florence.

Newland sat next to the Countess. He did this so that everybody at the opera could see him.

"We used to play together when we were children," said Countess Olenska. "You were a horrible boy. You kissed me once behind a door, but I was in love with your cousin Vandie Newland, who never looked at me." She looked around the opera house and said, "Yes. Being here brings back all the old memories. I can imagine everybody here in children's clothing just like long ago."

Newland was shocked by the flippant way [2] she referred to New York high society, which, at that very moment, was judging her. "You've been away a very long time," he said.

"Oh yes," replied the lady. "Centuries and centuries; so long that I feel as if I'm dead and buried, and this dear old place is heaven."

To Newland, her way of speaking seemed very strange. He didn't like her tone — it was too European, too subtle. He thanked God that he was an honest New Yorker and that he was about to marry one of his own kind.

1. **amber** : a hard yellow-brown substance used for making jewellery.
2. **flippant way** : showing that she's not taking New York as seriously as Newland thinks she should.

The text and **beyond**

1 Comprehension check

Answer the questions below.

1 What is Newland's attitude towards his job?
2 Who is Countess Olenska?
3 What does Newland think about the fact that Mrs Welland helps Countess Olenska?
4 How did Countess Olenska end up in New York?
5 Why was she "unfortunate"?
6 What is the suspicious part of her story, according to Larry Lefferts?
7 Why did Newland go over to Mrs Mingott's box?
8 How did Ellen shock New York when she was a child?
9 Who was Mrs Medora Manson?
10 What was Ellen's husband like?
11 What did Ellen remember about Newland when he was a little boy?
12 How did Ellen speak about New York?

2 Speaking

Obviously things have changed a lot since the 1870s. Or have they? What do men want in a wife today in your country? Make a list with your friends and then write down your ideas in a report. Use the list below to help you.

- attractive physical appearance
- a good job
- a good personality
- similar hobbies and leisure interests
- the same educational level
- the same ethnic or cultural background
- somebody they can feel superior to
- *your own idea ...*

3 Tragic innocence

Edith Wharton felt nostalgia for some of the old ways of her youth in New York. But she did not like the way in which young women were brought up. For questions 1-12, read about what Edith knew before she got married, and think of the word which best fits each space. Use only one word in each space. There is an example at the beginning (0).

Newland considers (**0**) with.... pleasure the innocence of May Welland. He is happy to see her blush as she listens to the love song from the opera *Faust*. This means that May's mother has (**1**) .... a good job: she has presented her daughter to society properly dressed and educated on the outside and properly empty on the inside. This picture of young feminine innocence was not the product of Edith Wharton's imagination; (**2**) .... was the product of her (**3**) .... experience. Edith herself wrote how her mother never offered her (**4**) .... information at all about the intimate relationships between husbands and wives. In fact, shortly before her marriage, the young Edith Wharton was desperate. Finally, she (**5**) .... the courage to go to her mother and ask, "What is marriage really (**6**) ....?" Her mother, who obviously did not approve of this (**7**) .... of question responded, "I have never heard (**8**) .... a ridiculous question!" Normally, the young Edith (**9**) .... have stopped there — she was very afraid of her severe mother — but she was desperate, and so she continued, "I'm afraid, Mamma —I want to know what will happen to me."

Her mother was silent for some time, and then, making a great effort, said, "You've seen enough pictures and statues (**10**) .... your life. Haven't you noticed that men are (**11**) .... differently from women?" Edith answered, "Yes," but without really understanding her mother's point. "Well, then?" concluded her mother. But Edith still looked at her without understanding. Finally, her mother ended the conversation by saying "Then for heaven's sake don't ask me any more silly questions. You can't be as stupid as you pretend."

But Edith was not pretending, and later in life she came to believe that this ignorance of the intimate relationships between husbands and wives caused her great harm and "misdirected" her life. Also, she felt that this ignorance often (**12**) .... tragic effects on young women.

Before you read

1 Pronunciation

Find the word in the box that rhymes with each word below. There are seven words that you do not need to use.

mean	**bread**	**warm**	**skull**	**turtle**	**ice**
term	**rocks**	**piece**	**beyond**	**shuttle**	**pull**
list	**teased**	**sung**	**speed**	**wrong**	

1 subtle ...shuttle...
2 kissed ....
3 dull ....
4 tongue ....
5 scene ....
6 bead ....
7 Nice ....
8 firm ....
9 box ....
10 blonde ....

2 Prediction

New York society is not happy to see Ellen.
How do you think the families of Newland and May will react? Briefly explain your answer.

Will they:

A try to convince Ellen to return to Europe?
B try to convince Ellen not to be seen in public?
C use all their social influence to convince people to meet her?
D *your own idea ...*

3 Reading pictures

Look at the picture on page 23 of Ellen.

- Where is she?
- What is she doing?

CHAPTER **TWO**

Invitations

That evening, Mr Julius Beaufort and his wife Regina gave a ball. They had a splendid house, and their annual[1] ball was always a great event in New York society. Regina Beaufort was from the Dallas family of South Carolina, and Mrs Mingott was her aunt, but her husband Julius – though rich and charming – was a mystery. He was a very successful banker and he claimed to be English, but nobody knew his family, and his behavior wasn't at all what New York generally approved of: he had a mistress called Fanny Ring. Everybody – including his wife – knew about this mistress, but nobody discussed her openly. Beaufort kept another house and a carriage for Fanny Ring. When they whispered to each other, the members of New York high society said that it was a terrible scandal. Nevertheless, they continued to accept Beaufort's invitations.

When Newland and May became engaged that afternoon they had decided not to make a public announcement for some time,

1. **annual** : happening or done once every year.

but now Newland told May that he wanted to announce it at once, at the Beauforts' ball. They did so, and they were both very happy. Newland was sure that it was the right thing to do. In announcing his engagement, he was announcing the connection of two important New York families: the Archers and the Wellands. If people wanted to gossip about Countess Olenska, now they would be going against two powerful families, not just one.

*

A few days later, Sillerton Jackson came to dinner at Newland's house, where he lived with his mother, Mrs Archer, and his sister Janey. When dinner was over, the ladies went to the drawing room. Newland and Mr Jackson stayed in the dining room to enjoy their brandy and cigars.

"Have you ever met Count Olenski?" asked Newland.

"Yes, once," Mr Jackson replied. "He's a very handsome fellow. He collects china – and women. I hear he'll pay any price for them."

"It's a good thing she left him. He sounds horrible."

"There are rumors about the Countess," said Mr Jackson.

"I know," said Newland impatiently. "Olenski's secretary helped her to run away from him. I hear the Count was almost keeping her a prisoner in that house, while he went out and spent his time with prostitutes [1]. I admire the secretary for helping her. Any gentleman in his position would have done the same!"

Mr Jackson smiled and looked at his cigar. "I hear he was still helping her a year later," he said. "They were seen together in Lausanne, Switzerland. They were living together."

1. **prostitutes** : sb who are paid to have sex with people.

Newland hesitated [1] for a moment and then said, "And why not? Just because she made a mistake in marriage, why should her whole life be over?"

"They say she wants to get a divorce," said Mr Jackson.

"Good idea!" Newland replied. "Women ought to be free! As free as we are!"

Mr Jackson poured himself more brandy and said, "Apparently Count Olenski agrees with you. He never made any effort to get her back."

When Newland had spoken so warmly about freedom for women, he had been entirely sincere, but in fact he wasn't quite so radical [2] as his words suggested. Privately, he thought that a "nice" woman would never take advantage of such freedom, even if it were given to her.

*

A few days later a terrible thing happened. Mrs Lovell Mingott sent invitations out, asking people to a dinner party at her house "to meet Countess Olenska". Of those invited, only three accepted. Everyone else said that they couldn't come. They gave no reason: they simply refused the invitation. This was an insult [3]. Clearly New York high society refused "to meet Countess Olenska".

"This is awful!" cried Mrs Archer when Newland told her about it. "We can't tolerate this. Our family is now linked to theirs through your engagement to dear May. We must do something. I know! Let's go and visit cousin Henry and see what he has to say about it!"

1. **hesitated** : to be slow to speak or act because you feel uncertain or nervous.
2. **radical** : in favour of thorough and complete political or social change.
3. **insult** : a remark or an action that is said or done to offend sb.

Newland agreed. It was an excellent idea. Mr and Mrs van der Luyden were at the very top of New York society. Their family was old, not just by New York standards but also by European standards, and they had several aristocratic [1] relatives in Europe. If Mr and Mrs van der Luyden accepted Countess Olenska, the rest of New York society would have to accept her too.

Newland and his mother went to see the van der Luydens that evening. When Mrs Archer had explained all about Countess Olenska and the refusals of Mrs Mingott's dinner invitations, the van der Luydens looked very worried indeed.

"Well," said cousin Henry after a while. "The Wellands and the Mingotts are connected to our family now, so we must do something about this. It is the principle of the thing that worries me: if an established New York family supports one of its members in her misfortune, the rest of society ought to accept that and support her too." He looked at his wife.

"My wife's cousin the Duke of St Austrey is coming to stay with us next week," said Mr van der Luyden. "We'll give a little dinner party for him and invite the Countess."

"Thank you so much!" said Newland. "That is sure to solve the problem."

After the Archers had left, Mrs van der Luyden took her elegant carriage and went to visit Mrs Mingott. Two hours later, everybody knew that Mrs van der Luyden's carriage had been seen outside Mrs Mingott's door. By the next morning they also knew that the purpose of Mrs van der Luyden's visit had been to invite Countess Olenska to a dinner party for the Duke of St Austrey.

A week later, as he sat in the van der Luyden's drawing room

1. **aristocratic** : belonging to or typical of the highest social class.

waiting for Ellen to arrive, Newland thought about her history and her strange, unconventional education. She walked into the drawing room half an hour late, wearing one glove and fastening a bracelet around her wrist, but she didn't look hurried or anxious. On the contrary, she was quite serene [1].

As Henry van der Luyden introduced her to his wife's cousin, the Duke of St Austrey, it was clear that he thought he was doing her a great honor, but she didn't seem to think so. Apparently she had already met the Duke in Nice. After dinner, the Duke sat beside her on the sofa in the drawing room, but after twenty minutes of conversation Ellen left him and crossed the room to sit beside Newland. It wasn't traditional in New York for a lady to leave the company of one gentleman and seek that of another, but Ellen seemed unaware of this.

"You know the Duke already?" asked Newland, as she sat down beside him.

"Yes. He likes to gamble [2]. He was often at our house in Nice. I think he's the dullest man I ever met, but people here seem to admire him."

Newland was a little shocked, but he laughed.

"Tell me all about May," said Ellen. "Are you very much in love?"

"As much as a man can be," Newland replied.

"Do you think there's a limit?"

"If there is, I haven't found it."

She smiled with real pleasure. "Then it really is a romance? It wasn't arranged by your families?"

1. **serene** : calm and peaceful.
2. **gamble** : risk money on a game.

"We don't allow our families to arrange our marriages here," said Newland.

She blushed. "Ah, yes!" she said. "I had forgotten that everything here is good that was bad where I've come from." She looked down at her hands, and her lips trembled.

"I'm so sorry," he said. "You are among friends here, you know."

"Yes, I know. Look! May has arrived. You'll want to hurry away and be with her."

The drawing room was filling up with after-dinner guests. May was with her mother. She was wearing a beautiful white and silver dress. She looked like the goddess Diana [1].

"She's already surrounded by other men," said Newland. "Look! The Duke is being introduced to her."

"Then stay with me a little longer," said Ellen quietly.

"Yes," replied Newland.

Mr van der Luyden came up and introduced Ellen to another gentleman. Newland stood up. Ellen turned to him and said, "I'll see you tomorrow after five, then."

"Yes, after five," Newland replied, though he was confused [2]. It was the first time she had mentioned an appointment.

1. **goddess Diana** : in Roman mythology, Diana was the goddess of the hunt.
2. **confused** : unable to think clearly or to understand what is happening or what sb is saying.

The text and **beyond**

1 Comprehension check

Match the phrases in column A with those in column B to make true sentences. There are four phrases in column B that you do not need to use.

A

1 ☐ May and Newland announced their engagement
2 ☐ Newland thought Count Olenski was horrible
3 ☐ Sillerton Jackson doubted the good intentions of Countess Olenska's secretary
4 ☐ New York society did not accept Mrs Lovell Mingott's invitations
5 ☐ Henry van der Luyden invited Countess Olenska to his house
6 ☐ In the end, New York society decided to accept Countess Olenska
7 ☐ The Duke came over and talked with Ellen
8 ☐ Ellen shocked Newland
9 ☐ Ellen got embarrassed
10 ☐ Newland was confused when Ellen confirmed their appointment the next day

B

A because she was invited to Mr and Mrs van der Luyden's dinner party.
B because they did not wish to meet Countess Olenska.
C because he was so happy that he was marrying May.
D because he was good friends with Count Olenski.
E because he thought the most important families should always help each other.
F because she hadn't even asked him yet.
G because she called the Duke boring.
H because he stayed with her for a long time after she had escaped from her husband.
I because she confused European traditions with American ones.
J because they wanted to protect May's cousin.
K because he did not let his wife leave the house while he went out with other women.
L because they were so happy about it.
M because he didn't think that she wanted to see him.
N because he already knew her.

FCE 2 The original ball

For questions 1-12, read the text below about the most famous ball in New York City. Use the word given in capitals at the end of each line to form a word that fits in the space in the same line. There is an example at the beginning (0).

The most (0) ..famous.......... ball of the late 19th century	**FAME**
was the Patriarchs' Ball, which began in 1873. The main	
purpose of this social (1) .... was, of course, to	**GATHER**
dance. Almost since its (2) .... by the Dutch in	**FOUND**
1616, New York, which was then (3) .... as	**KNOW**
New Amsterdam, had been a city of parties.	
The upper classes, however, had another important reason	
for having balls: they needed an occasion to present their	
daughters in the most (4) .... society.	**FASHION**
Marriage, of course, was not just a question of love, it	
was also a question of alliances of power and money	
among New York's (5) .... families.	**PRIVILEGE**
These balls (6) .... showed off the wealth of	**CERTAIN**
New York's top families.	
They always had the best food, wine, service, music and	
floral (7)	**DECORATE**
One of these dinners was (8) .... : the table	**BELIEVE**
had live swans swimming around the green islands in	
a miniature lake about thirty feet long in the center of	
the table!	
Money, however, was not the only point.	
As the (9) .... of the Patriarchs' Ball, Ward	**ORGANIZE**
McAllister said, "A fortune of a million is only	
respectable poverty." This funny and	
(10) .... remark does have a real meaning: the	**SNOB**
ruling families wanted their daughters to marry only	
(11) .... young men from other ruling families	**REFINE**
and not the sons of the many businessmen who had	
made fortunes during and after the (12)	**AMERICA**
Civil War.	
The Patriarchs' Ball was another way of trying to keep	
the power of New York in the hands of the families who	
had ruled it for the last 200 years.	

3 Odd word out

Choose the odd word out, and then say why it does not belong with the other three.

0 theater / ballet / opera / soprano

soprano is a kind of singer, the other three are types of performance.

1 secretary / banker / lawyer / business

2 European / Polish / American / New Yorker

3 glove / bracelet / diamonds / beads

4 stare / see / watch / look

5 bracelet / hairstyle / dress / glove

6 rumor / mistress / gossip / news

7 marriage / relatives / engagement / husband

Before you read

1 Listening

Listen to the beginning of Chapter Three. You will hear about Countess Olenska's home. Then answer the following questions.

1 Who were Ellen's neighbors?

2 Where was Ellen's maid from?

3 Did Newland think that May would be happy that he had gone to see Ellen?

4 What was Ellen's drawing room like?

5 How many roses did Ellen have in her drawing room?

6 Why was this unusual in New York?

7 Who would decorate Newland and May's house?

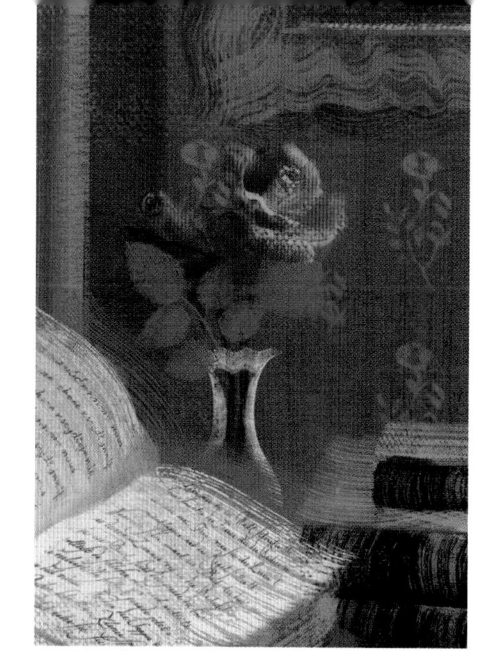

CHAPTER **THREE**

Samarkand[1] and New York

Countess Olenska lived in a bohemian part of the city, the kind of place where artists and writers live. At five-thirty the next day, Newland arrived at the house. An Italian maid showed him into the drawing room. She said that the Countess was out but that she'd probably be home soon.

Newland had spent the afternoon with May and her mother, going to visit friends and relatives. He had hardly had a moment to speak to May alone, so he hadn't told her about Countess Olenska's request — her command — that he should visit her after five, but he knew that May would approve: she was always asking him to be kind to her cousin. After all, it had been in part to protect the Countess that he and May had announced their engagement sooner than they had planned. If Countess Olenska hadn't come to New York, he would still have been a free man.

1. **Samarkand** : the second-largest city in Uzbekistan. Here it refers to Ellen's house which seems like a romantic and exotic place to Newland.

The drawing room was beautiful and unconventional. It smelled of spices. Several modern paintings in old frames hung on the red walls. French novels lay on the table, and beside them stood a vase with two roses in it. In New York, no one ever left books in the drawing room, and no one ever bought less than a dozen roses. He tried to imagine the drawing room in his future home with May. May and her mother would decide exactly how it should look, and it would be completely conventional, nothing like this room.

Hearing a carriage arrive at the door, he went to the window and looked out. There he saw Countess Olenska getting out of Beaufort's carriage, followed by Beaufort himself. Beaufort kissed her hand and got back into the carriage.

"Ah!" cried Countess Olenska, coming into the drawing room. "How do you like my house?"

"It's lovely," said Newland.

"I like it. I'm glad it's here in New York – in my own country and my own town. And I'm glad I live alone in it."

"Do you like being alone?" asked Newland.

"Yes, as long as my friends visit me so that I don't feel lonely." She sat down near the fire and said, "This is the time of day I like best."

"I was afraid you'd forgotten the time. Beaufort can be very charming."

"Mr Beaufort took me to see some houses. My family don't like this one. I don't know why. This street is respectable."

"But it isn't fashionable," Newland replied.

"Is that so important?" she asked with a laugh, then she added, "but I want to do what you all do. I want to feel cared for and safe."

"New York is terribly safe," he said ironically[1].

1. **ironically** : Newland means that New York is dull and predictable.

"Yes. I feel that," she replied. She hadn't noticed his irony. She offered him a cigarette and lit one herself. "You must help me. You must tell me what I should and shouldn't do."

He wanted to say, "Don't drive around with Beaufort". But that was New York advice, and he didn't feel as if he were in New York here. This seemed more like a drawing room in Samarkand. "There are plenty of people to tell you what to do," he said.

"Yes — my aunts and my grandmother. They've all been so kind. But they don't want to hear anything unpleasant. I tried to talk to them, but my Aunt Augusta told me it's better not to discuss these things. Doesn't anyone here want to know the truth, Mr Archer? I feel so lonely living among all these kind people who want me to pretend!" She began to cry.

"Countess Olenska! Ellen! Don't cry!" he said, touching one of her hands.

"Does no one cry here, either?" she asked, moving her hand to wipe her tears away.

Just then the Italian maid came in and announced the Duke of St Austrey. Newland rose to his feet. "I'd better go," he said.

Out in the street, he felt that he was in New York once more. He stopped at a florist's shop to send lilies of the valley [1] to May. He did this every morning, but today he had forgotten. Looking round the shop, he saw a vase full of yellow roses. He asked the florist to send them to Countess Olenska's address. He didn't sign the card.

*

The next day Newland went to see May. "Thank you for my lilies of the valley! They smell so lovely. It's so good of you to remember to send them every day!"

1. **lilies of the valley** : small, intensely perfumed, white flowers.

"They were late yesterday," said Newland. "I didn't have time to send them in the morning. I sent your cousin some yellow roses at the same time. I hope that was the right thing to do."

"How kind of you! She had lunch with us today, but she didn't mention the roses. She said she'd received flowers from Mr Beaufort and from Mr van der Luyden. She seemed so pleased."

Newland was annoyed that his own flowers had not been mentioned, even though he had failed to sign the card. Impulsively [1], he said, "May, let's get married sooner than we planned. Why wait?"

"Well, it's usual to wait a little while. Most New York couples are engaged for a year or two."

"Why can't we be different?"

"Oh, Newland! I love you so much! You're so original!"

"Original!" he cried. "On the contrary, we're all like paper dolls [2], exactly the same: we do the same things; we say the same things." He had an irritating sense that May was playing the part of a young woman in love, saying all the things such young women were supposed to say.

"Mother wouldn't like it if we were different," said May. She looked a little bored and irritated, but then she smiled and said, "Oh! Did I tell you? I showed my engagement ring to Ellen. She thinks it's the most beautiful ring she's ever seen. She says there is nothing like it in Paris! I do love you, Newland, for being so artistic!"

1. **Impulsively** : acting suddenly without thinking carefully about what might happen because of what you are doing.
2. **paper dolls** : figures cut out from a folded piece of paper, so that each figure is exactly like the others.

The text and **beyond**

FCE 1 Comprehension check

For questions 1-6, choose the correct answer — A, B, C or D.

1 The vase with the roses and the books made Newland think about his future because

A ☐ they were unusual things that May would never have in their home.

B ☐ they were just the kind of things that he imagined for his future home with May.

C ☐ roses and flowers made him think of his wedding with May.

D ☐ he imagined explaining interesting books to May when she became his wife.

2 Ellen's family did not approve of her house because

A ☐ they didn't choose it for her.

B ☐ it was not in an area that was then popular with the important families.

C ☐ they did not like the people who lived near her.

D ☐ it was too far away from their homes.

3 Newland didn't tell Ellen about Beaufort's bad reputation because

A ☐ he did not know her well enough.

B ☐ he felt that her house was somehow not part of New York.

C ☐ he did not wish to hurt her feelings.

D ☐ he was certain that her family would explain such things to her.

4 Ellen wanted Newland's advice on

A ☐ her unpleasant situation with the count.

B ☐ how she should act in New York.

C ☐ how to meet more good people in New York.

D ☐ how to find fashionable houses in New York.

5 Ellen felt lonely because

A ☐ her family never came to visit her in her new house.

B ☐ she could not understand the ways of New York society.

C ☐ nobody was kind to her.

D ☐ she could not talk with anybody about her problems with the Count.

6 May told Newland what Ellen thought of her engagement ring because
- A ☐ she wanted to change the subject of their conversation.
- B ☐ she was so proud of what Ellen told her.
- C ☐ she wanted him to know how important Ellen was to her.
- D ☐ she thought he would be happy that a sophisticated European woman liked it.

2 Summary

Number the paragraphs in the right order to make a summary of chapters One to Three. One has been done for you as an example.

- A ☐ However, a few days later, New York society showed its disapproval of Ellen by refusing an invitation to a party for her.
- B ☐ The next day Newland went to see May to ask her if they could marry sooner. Clearly, Ellen had had a big effect on him.
- C ☐ The van der Luydens were shocked when they heard about Ellen. They decided to invite her to a party in honor of the Duke of St Austrey.
- D ☐ Suddenly, one of his friends saw a new arrival in May's box. It was her cousin Ellen, now known as Countess Olenska.
- E ☐ A week later, Ellen arrived at the party. She talked to the Duke. She also talked with Newland and invited him to her house the next day.
- F [1] One evening a young man named Newland Archer was at the New York Academy of Music. But he didn't watch the singer, he watched his young fiancée, May Welland.
- G ☐ When he arrived, May's mother introduced him to Countess Olenska. She seemed very European to Newland.
- H ☐ Newland's friends began to gossip about Ellen. Newland got up and went over to May to show his support for her and her family.
- I ☐ Afterwards, everybody went to a ball. It was here that May and Newland announced their engagement. Now their two families were connected, and could better protect Ellen.
- J ☐ Newland arrived at Ellen's, but she was not there yet. He was very impressed by her house. Finally, she arrived. Ellen asked him to help her understand the ways of New York.
- K ☐ Newland's mother was horrified. She decided to ask her socially important cousin Henry van der Luyden for help.

3 Marriage now

Why do you think Newland all of a sudden wants to be married so soon? Choose one of the answers below or give your own. Justify your answer.

A Seeing Ellen's "unfashionable" house makes Newland want to do something different.

B He wants to begin his new life and begin educating May.

C He feels attracted to Ellen and wishes to avoid the temptation.

D Your idea.

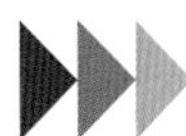

INTERNET PROJECT

To take a tour of Newland Archer's New York go to the Internet and go to www.blackcat-cideb.com or www.cideb.it. Insert the title or part of the title of the book into our search engine. Open the page to *The Age of Innocence*. Click on the Internet project link. Scroll down the page until you find the title of this book and click on the relevant link for this project. Then type in 'New York City' at 'Search this collection'. Download some pictures that you think are particularly interesting. Explain to your partner or the class why you like them.

An 1840 painting of **Five Points**, a slum area of Manhattan from 1820 to 1890. It was home to poor Irish immigrants in the 1840s and then African Americans after the Civil War (1861-65). It is the setting of Martin Scorsese's 2002 film ***Gangs of New York***.

New York in the 1870s

Imagine Manhattan before the Statue of Liberty, the subway system, the yellow taxi cabs and cars that speed through the streets, the Empire State Building and all the other skyscrapers that make up the amazing skyline. What's left? In the 1870s some of the typical brownstone residential buildings were already there: New Yorkers started building their homes in brownstone after the Great Fire of 1835 destroyed whole neighborhoods of wooden houses. Central Park was opened in 1873. And New York already had a large and rapidly growing population.

New York City was the capital of the United States until 1790. Its population was then only 33,131, but by 1850 it had grown to 696,490 and by 1900 it was close to 4,000,000. After the American Civil War (1861-5), the northern industrial States experienced a great economic boom. Many rich bankers and industrialists lived in Manhattan, where they and their families formed a very exclusive 'high society'. The most influential families were the 'old money' – a kind of American aristocracy who could trace their ancestors back to the original Dutch settlers of Manhattan Island in the seventeenth century, which was then called New Amsterdam.

In the 1880s, Samuel Ward McAllister (1827-95), a rich lawyer who attended New York high society balls and dinner parties, coined the phrase "the Four Hundred". According to him, this was the number of people in late-nineteenth-century New York who really mattered. The number was popularly supposed to be the capacity of Mrs William Backhouse Astor Jr's ballroom. The Astors were among the wealthiest and most influential families in New York.

In 1904, the American writer O. Henry published a book called *The Four Million*. The title was O. Henry's reaction to McAllister's phrase: in O. Henry's opinion, *every* human being in New York was worthy of notice. The Astors' son John Jacob Astor IV died on the *Titanic* in 1912 returning from his honeymoon with his second wife. He had divorced his first wife, which was a great scandal in New York high society at the time. John Jacob Astor IV was a character in the 1997 film *Titanic*. In fact, that film represented the whole social scale of New York in the early twentieth century, from the struggling immigrants in third class to the lavish [1] lifestyle of Astor – the wealthiest man on the ship. That social structure was already in place by the 1870s, and its ruling class is the subject of *The Age of Innocence*.

However, there were also great changes in both the look and the social make-up of Manhattan between the time of the setting of the novel in the 1870s and the time of its publication in 1920. After the

1. **lavish** : large in amount, or impressive, and usually expensive.

Mulberry Street, New York City, seen in a photograph from about 1900, was where many Italian immigrants lived after their arrival in America, and is still the center of Little Italy.

Civil War, there was a great migration of African Americans from the southern States to the industrial north. By 1916, New York City had the largest African American population in the United States. The immigrants found plenty of work as the city boomed, especially in the construction industry, as competing skyscrapers sprang up and the subway system was built. The Metropolitan Life Insurance Company Tower was constructed in 1909 and is 213 meters high. The Woolworth Building, constructed in 1913, is 241 meters high. These were the first skyscrapers and were followed by a rush of other taller buildings, such as the Chrysler Building and the Empire State Building in the 1930s.

The title of Wharton's novel, then, seems to suggest nostalgia for the old days before this dizzying growth. On the other hand, the portrait it paints of old New York is not at all "innocent": its ruling-class characters are guilty of narrow-mindedness, snobbery, and mental cruelty. F. Scott Fitzgerald, author of *The Great Gatsby*, famously said that an artist is someone who can have two contradictory ideas at the same time. In its condemnation[1] of the rigid old society of New York in the 1870s and its underlying nostalgia for that society's grace and simplicity, Wharton's novel vividly demonstrates Fitzgerald's idea.

1. **condemnation** : an expression of very strong disapproval.

1 Comprehension check

Answer the following questions.

1 When was New York City the capital of the United States?
2 What was the population of New York City in 1850? In 1900?
3 What was New York's first name?
4 Who were "the Four Hundred"?
5 Why did O'Henry call his book *The Four Million*?
6 How did Manhattan change from the 1870s to 1900s?
7 What did Edith Wharton like about the old society of New York in the 1870s?

Before you read

1 Listening

Listen to the beginning of Chapter Four. You will hear Newland talking with his boss, Mr Letterblair, about Ellen's possible divorce. Then answer the following questions.

1 Who asked Mr Letterblair for help concerning Ellen's divorce from the Count?
2 Why does Mr Letterblair want Newland to talk with Ellen?
3 Why did Mr Letterblair choose Newland for this particular job?
4 What is one reason Mr Letterblair gives to show that there is no real reason for Ellen to get a divorce?
5 What did the Count threaten to do if Ellen tried to divorce him?

CHAPTER **FOUR**

Marriage is Marriage

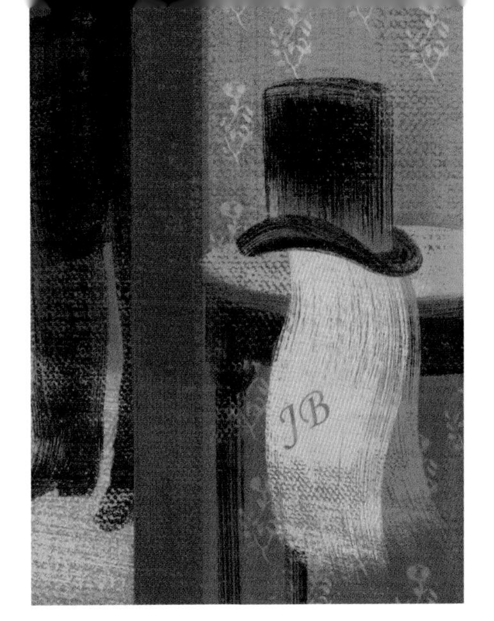

Two weeks later at the office, Newland was summoned by the head of the firm, Mr Letterblair, who said to him, "Mrs Manson Mingott sent for me yesterday. She says her granddaughter, Countess Olenska, wants to divorce her husband. Mrs Mingott gave me all the relevant papers and asked me to act for the family. She made it quite clear that the family don't want a divorce. It'd be a terrible scandal. I want you to go to the Countess and persuade her not to ask for a divorce."

"Can't someone else do it?" asked Newland. "The fact that I'm engaged to her cousin makes it rather difficult for me."

"It is precisely because you're engaged to her cousin that we're asking you to do it. Mrs Mingott asked for you. It is a very private matter and nobody else in the firm knows about it. Please take these papers and read them."

"Perhaps divorce is the best thing in this case. I can't promise to persuade the Countess not to divorce her husband unless I'm convinced[1] that it's the right thing to do."

"I don't understand you, Archer," said Mr Letterblair. "You're going to marry into her family. Do you want that family to be the

1. **convinced** : completely sure about sth.

subject of scandal?"

"No, of course not, but what I want is irrelevant[1]."

"The Countess is here; the Count is in Europe: the Atlantic is between them. The Count has already given her some of her money back[2]. He won't give any more. Why should he? The marriage agreement doesn't require him to do so, even in the case of divorce. Besides, they say she doesn't care about the money. If that is the case, she should just let things remain as they are."

Reluctantly, Newland took the papers to his office and read them. The last was a letter from the Count in which he threatened to create a scandal over the Countess's relationship with his secretary if she insisted upon a divorce. It was a very nasty letter. Newland felt a sudden compassion for the Countess. END The older ladies in New York society were very severe in their judgment of any woman who had a relationship outside marriage. They spoke of "that kind of woman" and showed no pity to her. They always pitied the man: they considered him a foolish creature who couldn't resist "that kind of woman"; the man was always the poor victim who had to be saved. Newland had never really questioned these beliefs, but now he suspected that in Europe things were less simple. There, he thought, it was possible for a good woman — an honest, sensitive woman — to fall into a relationship like that just because she was lonely and desperate[3].

The next day he went to see the Countess. As he entered the drawing room, Newland was unpleasantly surprised to see Julius Beaufort standing by the fire. "Why are you going to the van der Luydens' again?" Beaufort was asking the Countess as Newland came in. The Countess turned to Newland and offered him her

1. **irrelevant** : not important to or connected with a situation.
2. **some of her money back** : sum of money — in this case a large one — given by the woman's family to her husband when they marry.
3. **desperate** : you are in such a hopeless situation that you will do anything to change it.

JB

hand. Beaufort nodded at him and continued talking to the Countess. "You'll be bored to death there. Come to dinner with me instead. I've planned a dinner for you at Delmonico's [1]. I want to invite all the artists to meet you."

"Ah!" said the Countess. "That does tempt me! I haven't met any artists since I've been here."

"I know some painters," said Newland. "I'll introduce them to you if you like."

"Painters? Are there any painters in New York?" asked Beaufort with an ironic smile.

"Thank you," said the Countess to Newland, "but I meant dramatic artists: singers, actors, and musicians. My husband's house was always full of them." Then she turned to Beaufort, offered him her hand and said, "Goodnight. I have to discuss business with Mr Archer."

Beaufort kissed her hand, but he didn't look pleased. As he was leaving, he said, "Archer, if you can persuade the Countess to come to Delmonico's on Sunday, you can come too."

"So you care about art, Mr Archer?" asked the Countess, when they were alone.

"Oh yes. I always go to the exhibitions when I'm in London and Paris."

"I used to care about art too, but now I want a new life. I want to leave my old life behind and become an American just like everyone else."

"You'll never be like everyone else," said Newland.

"Don't say that. I hate to be different. I want a new start."

"I know. Mr Letterblair told me. In fact, he asked me to come to see you. That's why I'm here."

1. **Delmonico's** : a famous, expensive New York restaurant (see activity 2, p. 50).

"Ah!" said the Countess with a smile. "You mean I can talk to you? You'll help me?"

"Yes."

"Have you read the papers? Do you know about my life with my husband?"

"Yes," said Newland, blushing.

"Then you agree with me that I should get a divorce?"

"Well, I'm not sure. Your family don't want a divorce, and they've sent me, or rather my partner, Mr Letterblair, has sent me to explain to you their point of view."

"Have you read my husband's letter?"

"Yes."

"And isn't it horrible?"

Newland looked down. "Yes."

"I'm a Protestant," said the Countess. "Our Church doesn't forbid divorce in such cases."

"But," said Newland, "in his letter the Count threatens to cause a scandal if you insist upon divorce."

"What harm could that do me here?" asked the Countess.

"I'm afraid it could do you more harm here than anywhere. You see, New York society is a very small world, ruled by people with rather conventional ideas."

"Yes," she replied, and her lips trembled a little as she spoke. "That's what my family tell me. And do you agree with them?"

He stood up and walked to the fireplace. Staring into the fire, he said, "It's my business to explain how your family see these things. The Mingotts, the Wellands, the van der Luydens — all your friends and relations."

"Tell me what you think — sincerely," she said.

"Well, what would you gain from [1] a divorce?" he said, looking into the fire.

"My freedom! Is that nothing?"

He thought perhaps she wanted a divorce so that she could marry the secretary, and this thought made him angry. "But you're free now, aren't you?" he replied impatiently. "Is it worthwhile to make a scandal that will upset all your family and friends?"

"No," she said, and her voice sounded sad.

She stood up quickly, as if to indicate that their talk was over. "All right," she said. "I'll do as you say. I won't ask for a divorce."

Newland blushed and took both her hands in his. "I do want to help you," he said.

"I know. You do help me. Goodnight," she replied. He bent his head, kissed her hands, and left her.

*

"The Count has written to her," said Mrs Mingott. "He has asked her to go back. He has offered to give her back a lot of the money he received when they were married. Lovell and Augusta think she should go back, and I agree with them."

Newland was sitting with her in her drawing room drinking tea. "I'd rather see her dead," he said.

"Would you?" the old lady replied. "But here, you see, my granddaughter is at risk. A woman alone is always at risk. In Europe she has everything she could desire: jewels, fur coats, splendid houses, the company of artists and intellectuals. Marriage is still marriage, Mr Archer, and my granddaughter is still a wife."

1. **gain from** : get advantage from, benefit from.

The text and **beyond**

1 Comprehension check

Who said what and why? Match the quotes with the character who said them, and them match the quotes with the reason why they said them. You may match some characters with more than one quote. There is an example at the beginning.

Who

Julius Beaufort (B) Ellen (E) Mr Letterblair (L)
Newland (N) Mrs Manson Mingott (M)

What

A [L] [4] "It'd be a terrible scandal."
B [ ] [ ] "The Atlantic is between them."
C [ ] [ ] "You'll be bored to death there."
D [ ] [ ] "I know some painters."
E [ ] [ ] "And isn't it horrible?"
F [ ] [ ] "I hate to be different."
G [ ] [ ] "Marriage is still marriage."
H [ ] [ ] "New York society is a very small world, ruled by people with rather conventional ideas."
I [ ] [ ] "My freedom!"
J [ ] [ ] "I'm a Protestant."

Why

1 He/She wants to leave his/her European life behind.
2 He/She is saying why he/she would be ready to face a scandal.
3 He/She wants to convince the Countess to spend the evening with him/her.
4 He/She is explaining why divorce will not be accepted.
5 He/She is saying why Ellen should not divorce, even if he/she understands her point of view.
6 He/She is saying why there is no real need to divorce.
7 He/She is saying why he/she can have a divorce.
8 He/She is saying why the Countess shouldn't ask for a divorce.
9 He/She is jealous of the Countess's interest in Julius Beaufort.
10 He/She thinks Newland agrees with him/her.

2 Delmonico's – an agency of civilization

You are going to read an article about Delmonico's, America's first great restaurant. Seven sentences have been removed from the article. Choose from the sentences A-H the one which fits each gap (1-6). There is one extra sentence which you do not need to use.

The French chef Antoine Beauvilliers opened the first modern restaurant in 1782. (**0**) E....

However, America and New York City would have to wait until the arrival of two Swiss brothers to begin to enjoy the wonders of fine French food in an elegant restaurant. (**1**) .....

Fortunately for America, a Swiss-born ship captain named Giovanni Delmonico grew tired of sailing. He knew something about wines. He wrote to his brother Pietro, who made candies and pastries. They opened a shop in New York in 1827. (**2**) ..... They used the best and freshest ingredients, and kept their shop spotlessly clean – this was revolutionary. They were extremely successful.

In 1831, they opened the Restaurant Français, which served hot lunches to New York's businessmen. (**3**) .....

That same year, the two brothers called their nephew Lorenzo over from Europe. (**4**) .....

In 1862, Lorenzo hired the great French chef Charles Ranhofer. Now, Delmonico's, as their restaurant was known, had a 12-page menu. Also, if you couldn't find what you wanted on that menu, all you had to do was ask the chef and he would prepare it for you. The meals served to New York's elite were incredible. The feast presented in 1868 to the visiting celebrity, the English writer Charles Dickens, started with oysters, two soups, a cheese timbale, salmon and bass, filet mignon, stuffed lamb, braised lettuce, grilled tomatoes and then went on to the main course of more than 30 dishes! **(5)** .....

But Delmonico's also had another role in New York: it was the place where high society held its most important balls. **(6)** ..... Then, in 1870, Archibald Gracie King broke with tradition and organized the ball to present his daughter to society at Delmonico's. His ball with some 800 guests was a huge success. So, it was then that Delmonico's also became the center of New York society.

A This too was a huge hit.

B It is not hard to understand why one British magazine wrote that the two wonders of America were "Yosemite Valley and Delmonico's".

C They served wine, coffee, chocolate and French pastries.

D Of course, you could eat out, but you ate what they gave you and your health was always at risk.

E Soon his idea caught on all over Europe.

F Before 1870, the elite families organized these events at home.

G He prepared some of the most fabulous meals ever seen in the United States.

H He would turn their restaurant into a New York institution, and an "agency of civilization" in America.

3 Vocabulary

Complete the crossword puzzle. All the words come from chapters One to Four.

Across

4 maybe
6 reason
7 with light hair
9 done each year
11 part of a play
12 have an odor
13 say that you will do something bad to somebody
14 extremely surprised
15 the opposite of 'hated'

Down

1 shake
2 eccentric and artistic
3 trendy
5 alternative, different from the usual
7 when your face becomes redder than usual because you are embarrassed
8 past participle of 'do'
10 your sister's or brother's daughter
12 accepted measures of comparison, criteria

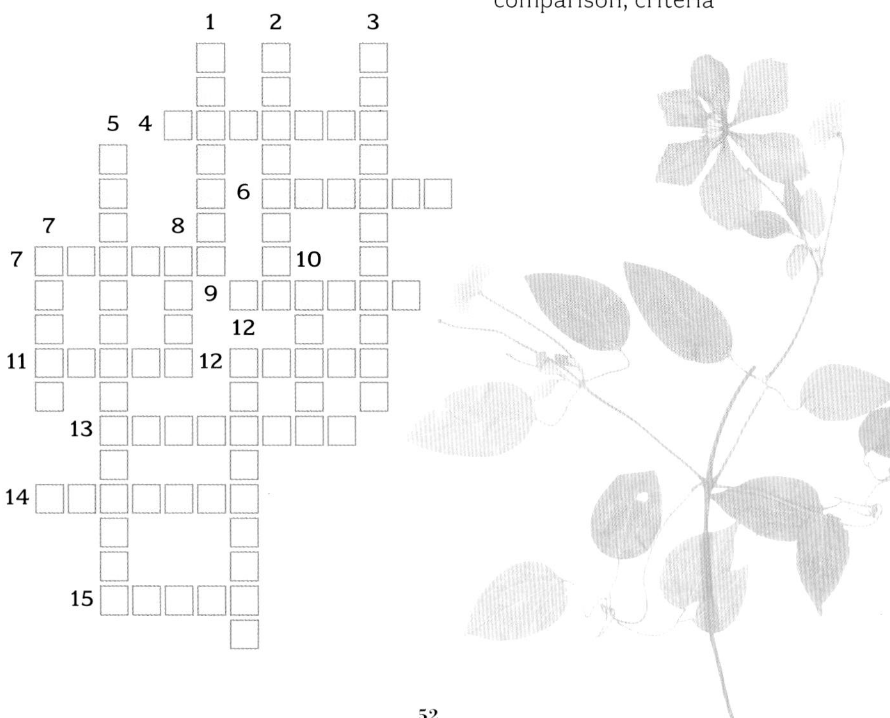

Before you read

1 Prediction

Here are two questions about events in the following chapter. With your friends guess the answer.
Present and explain your answers to one another.

1 May will say no when Newland asks her to get married earlier than planned. Why?

- **A** It will ruin the complicated plans that have been made for the wedding.
- **B** She is beginning to have doubts about Newland and wants time to think.
- **C** She thinks Newland loves another woman and wants him to be certain of his choice.
- **D** *Your own idea ...*

2 Newland tells Ellen he loves her. What will her reaction be?

- **A** She will threaten him to tell May if he does not stop acting so stupidly.
- **B** She will offer to be his mistress.
- **C** She will get angry and tell him to never talk to her again.

2 Reading pictures

Look at the picture on page 57.

- Who is in the picture?
- What are they doing?
- What emotion is illustrated by this picture?

Look at the picture on page 61 of Newland and Ellen.

- What is Newland doing? What is Ellen doing?
- What do you think is going to happen next?

CHAPTER **FIVE**

The Wedding

In March, May went to St Augustine in Florida with her parents for a month. Newland went down to see her. He returned two days later and went straight to Countess Olenska.

"I went to St Augustine to ask May to marry me after Easter, instead of waiting for another year, but she wouldn't agree."

"Why not?" asked the Countess, lighting a cigarette.

"She wants to give me time."

"For what?"

"She thinks my impatience is a bad sign. She thinks I want to marry her soon to get away from someone else I love more."

"She wants to give you time to give her up for another woman? That's very noble of her."

"Yes, but it's ridiculous."

"Why? Because you don't love anyone else?"

"Because I don't plan to marry anyone else," said Newland.

"Ah!" There was a long silence. "And is there another woman?"

"Yes. May's right. There is someone else." He put his hand on hers.

She stood up quickly and walked to the other side of the room. "Don't do that!" she cried.

Newland stood up. "I'll never do anything to offend you," he said. "But you're the woman I would've married if it had been possible for either of us."

"Possible for either of us?" she cried in amazement. "But you're the one who made it impossible! You made me give up the idea of getting a divorce! You told me to save my family from scandal! And because my family was soon to be your family — for May's sake and for yours — I did what you told me to do."

"I thought —" began Newland. "I thought you were afraid of the scandal if the accusations [1] in your husband's letter were made public."

"I had nothing to fear from that letter! All I feared was bringing scandal on the family — on you and May."

"Good God!" he cried, putting his face in his hands. Then he heard her crying by the fire. He went over to her and said, "Ellen. I'm still free, and you can get a divorce. We can be happy!" He took her in his arms and kissed her wet face.

"No, we can't," she said. "You're engaged to May, and I'm married."

"How can I marry May after this?" he cried.

"You must. You don't understand how you've changed things for me. I didn't realize that people disapproved of me. I didn't know that they all refused Granny's [2] invitation to a dinner to

1. **accusations** : a claim that sb has done sth illegal or wrong.
2. **Granny** : an affectionate abbreviation of "grandmother".

meet me. Later, one day when she was angry with me, Granny told me everything. She said that you went to the van der Luyden's and asked them to invite me to dinner. She told me that you and May announced your engagement early so that I would have the protection of two families, not just one. I hadn't understood anything. Everyone seemed so kind. But no one was as kind as you. You explained to me why it was bad to ask for a divorce, and you were right. You showed me that it is wrong to find happiness by making other people suffer."

As she spoke, he sat beside her, looking at the tip of her satin shoe sticking out from under her dress. Suddenly he fell to his knees on the floor and kissed the shoe. "I don't want to go back to my old way of thinking," she said. "Don't you see? I can't love you unless I give you up."

Newland replied bitterly: "And Beaufort? Is he going to replace me?"

He expected her to be angry, but she just went a little paler. "No," she said.

"Why not?" cried Newland. "You say that you're lonely."

"I won't be lonely anymore," she said. "Now that I know you love me."

Just then, the maid came in and handed a telegram to Ellen. She opened it, read it, and handed it to Newland.

It was from St Augustine:

Papa and Mama have agreed that Newland and I can marry after Easter. I'm sending a telegram to Newland now. I'm so happy!
Love, May.

Newland and May were married the first Saturday after Easter. As he went through the ceremony and received everyone's congratulations, Newland felt cold and empty. They went on their honeymoon [1] to all the usual destinations: London, Paris, Florence, Rome. Every hour of every day he was with May, listening to her innocent chatter [2]. Often she irritated him. Her ideas were so conventional that — away from New York — she seemed narrow-minded and dull.

They went home in July. There, Newland felt more at ease with May. She was one of the prettiest and most popular young wives in New York, and he was proud of her. A year passed. Newland now thought of his passion for Ellen as a moment of madness. How could he ever have thought of marrying her? It was ridiculous. Now she lived on in his memory only as a strange eccentric [3] fascination [4].

In August, New York's best families went to Newport. Newland and May were no exception. They were there with their families — Newland's mother and sister; May's parents, aunt, and uncle. One day, they went to visit May's grandmother Mrs Manson Mingott. She'd just arrived from the city and was staying in her house by the sea. They found her sitting in a garden chair beneath a big tree. Newland and May sat beside her, and Mrs Mingott rang the bell for tea.

"How nice to see you, Newland!" cried the old lady. "Will you be here for the whole month of August?"

"May will," Newland replied. "I must return to the city for

1. **honeymoon** : the vacation a newly married couple takes together just after the wedding.
2. **chatter** : conversation about unimportant things.
3. **eccentric** : sb who behaves in strange or unusual ways.
4. **fascination** : a very strong attraction.

business reasons from time to time."

"Ah! Business! Many husbands find it impossible to join their wives here except at the weekends. Marriage is one long sacrifice, as I often say to Ellen."

Newland's heart seemed to stop at the mention of her name.

"Julius Beaufort is here now, but poor Regina has had to spend most of the month here alone. Julius's business keeps him in the city most of the time." Here Mrs Mingott raised an ironic eyebrow, because everyone knew that the business that kept Beaufort in the city was a lady named Fanny Ring. "From what I hear," continued Mrs Mingott, "his business isn't going very well. Apparently he invested in railways and lost a lot of money. That doesn't stop him spending money here, though. He came this morning to see Ellen, and he was telling me all about his new racehorses."

"Is Ellen here?" asked May.

Newland didn't breathe as he waited for the reply. "Well, actually she's staying with the Blenkers in Portsmouth, but she has come to see me for the day," said Mrs Mingott, then she cried out, "Ellen! Ellen!"

The maid came out of the house and said, "The Countess has gone for a walk by the sea, madam."

"Newland, will you go and find her?" asked Mrs Mingott.

"Certainly."

As Newland walked through the woods towards the sea, his heart beat fast.

He'd heard her name mentioned often enough in the eighteen months since he'd last seen her. He even knew the main events of her life. She'd spent the previous summer in Newport, but, in the autumn, she'd moved to Washington. Hearing about her had never disturbed him before, but now that he was going to see her again she became once more a warm living presence for him, and he remembered the lovely drawing room with the red walls...

Newland came out of the woods and looked down at the sea. There was a lighthouse and a pier [1]. The sun was sinking and the whole scene was bathed in golden light. Ellen stood on the pier, looking out to sea. Beyond her, a sailboat was crossing the bay. Newland thought, "If she turns round before the sailboat passes behind the lighthouse, I'll go to her." He watched her intently [2], but she didn't move. She stood perfectly still, looking out to sea. He waited until the boat was well past the lighthouse, then he turned and walked back to the house.

As he and May drove away from Mrs Mingott's, May said, "I'm sorry that Ellen wasn't there. She doesn't seem to care about her old friends anymore. I suppose she moved to Washington because New York bores her. And now, instead of staying with Granny in Newport, she's staying with the Blenkers in Portsmouth. Perhaps, after all, Ellen would be happier with her husband."

"I've never heard you say anything cruel before," replied Newland.

"Why cruel?" asked May in surprise.

"Life with her husband was hell. Do you think she'd be happier in hell?"

"Well, she shouldn't have married a foreigner, then," said May.

1. **pier** : a wooden walkway that goes out over the sea.
2. **intently** : showing strong interest and attention.

The text and **beyond**

1 Comprehension check

Say whether the following statements are true (T) or false (F), and then correct the false ones. If you do not have enough information to be certain, choose possible (P).

		T	F	P
1	Newland's love for May made him want to marry her as soon as possible.			
2	May did not believe that Newland's love for her was the cause of his desire to marry earlier.			
3	According to Ellen, it was her fault if she could not marry Newland.			
4	Beaufort will keep Ellen company now that Newland is going to marry.			
5	May knows that Newland is in love with Ellen.			
6	Newland accepted his marriage to May once they were away from New York.			
7	After his marriage, Newland did not know anything more about Ellen.			
8	Julius Beaufort began making large profits from the railways.			
9	Ellen's presence in Newport had no effect on Newland.			
10	May thinks that Ellen should divorce.			

"Well, she shouldn't have married a foreigner, then"

Look at the meaning of *should*

*You **shouldn't talk** with strangers = It is not a good idea to talk with strangers.*

*American women **should marry** American men.*
= It is a good thing that American women marry American men.

In other words, we can use ***should*** to express our opinions about things. To express an opinion about things that happened in the past we use ***should have*** or ***shouldn't have*** + **PAST PARTICIPLE.** (過去分詞)

*Ellen **shouldn't have married** a foreigner*
= It was not a good idea for Ellen to have married a foreigner.
= She made a mistake when she married a foreigner.

Ellen ***should have married*** *an American.*
= It would have been better if Ellen had married an American.

2 *Should have or shouldn't have*

Fill in the blanks below using one of the verbs in the box and *should have* or *shouldn't have*. The first has been done for you as an example.

accept live so long choose spend the afternoon
marry show more respect ~~bring~~ advise

0 Ellen Olenska was in Mrs Mingott's box at the Academy of Music. Larry Lefferts says, "They shouldn't have brought her to the opera."

1 Mr and Mrs van der Luyden have heard that hardly anybody is coming to Mrs Lovell Mingott's dinner party. They say, 'People .... the invitation.'

2 Newland sees that Ellen was with Julius Beaufort. He thinks to himself, 'She .... with a married man.'

3 The New Yorkers see the young Ellen dressed in bright colors and beads after her parents' death. They say among themselves, "She .... in Europe."

4 The New Yorkers hear about all the trouble Ellen has with her Polish husband. They say among themselves, 'She .... an American.'

5 Newland tells Ellen that he loves her and that he would like to marry her. Ellen says, 'You .... me against divorce.'

6 Ellen seemed rather flippant and ironic when she talked about New York society. Newland, considering her attitude the next day, says to himself, "She .... to New York families."

7 Ellen's family learns where she is going to live. They say among themselves, "She .... a house in a bohemian neighborhood."

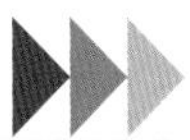 **INTERNET** PROJECT 

Newport: America's first resort

Newport is a town on Aquidneck Island. It has been a resort area for almost 300 years. However, its period of splendor was from 1870 to 1917, more or less the period covered in *The Age of Innocence*. It was then that America's wealthiest families built fabulous mansions on the island for their summer holidays. These buildings can still be admired today.
To find out more about the history of this fascinating town, go to the Internet and go to www.blackcat-cideb.com or www.cideb.it. Insert the title or part of the title of the book into our search engine. Open the page to *The Age of Innocence*. Click on the Internet project link. Scroll down the page until you find the title of this book and click on the relevant link for this project.

Prepare a short report to present to your friends. Use these questions to help you.

- How did Newport become a religious haven?
- What role did Newport play in the slave trade?
- What did the upper, middle and working classes do to enjoy themselves?
- What famous people spent time at Newport?

Download some of the old pictures. Print them and explain them to your friends.

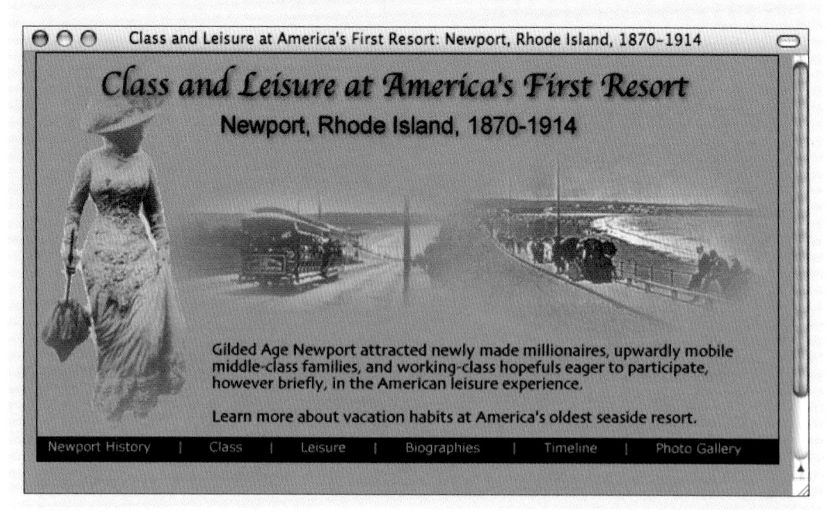

3 Speaking: American women should grow up

In 1919, just after the First World War, Edith Wharton wrote a book called *French Ways and Their Meaning*. Interestingly, she told her American readers that French women were superior to American women. She described American women as kindergarten children compared to their French counterparts who were "grown up". Besides the fact that French women generally cooked better, dressed better and were more sensual, they knew about real life. This was because the middle-class women in France were, explained Edith Wharton, very often their husbands' main partners in business. In this way, they learned to face the actual problems of life. American women on the other hand — despite all their freedom, talk and travel — just played at life.

These views come out clearly in *The Age of Innocence* where the best example of middle-class American woman is May, who, in some ways, is presented as a child. Ellen is like French women — she knows about the world in ways that are even shocking for Newland. Perhaps, however, the biggest difference between May and Ellen, is that Ellen is an interesting companion — May is simply somebody to teach.

With your friends discuss your opinions on women and men in your country. Write down a short report using the questions below to help you.

- Are there any general characteristics of women and men in your country?
- Are there any stereotypes?
- Are the women or men of another country sometimes considered superior in some way?

Before you read

1 Prediction

Newland does go to Boston to see Ellen. When he finds her he says, "Well, you see, it's no use. My 'business' in Boston was simply to find you. Come on. Let's go to lunch."

- What do you think Newland will say to Ellen during this lunch?
- What do you think her responses will be?

CHAPTER **SIX**

Boston

Mrs Archer said they'd all been invited by Professor and Mrs Emerson Sillerton to a party for the Blenkers. "It's a terrible bore, of course, but the Sillertons are related to Sillerton Jackson, so I suppose at least some of us will have to go."

"I'll go with Mother," said May.

"I'm afraid I can't go," said Newland. "I've arranged to go to a farm in the north to look at some horses."

Having said that he'd be gone all afternoon, it actually took him just an hour to drive up, see the horses, decide that he didn't want to buy them, and leave. The rest of the afternoon was free. He drove to the Blenkers' house near Portsmouth. He told himself that he didn't want to see Ellen, but he had a strong desire to see the house she was living in. He'd go there and look at the place. Then later he'd be able to imagine her eating breakfast there or walking in the garden. If he could do that, perhaps the world around him would feel less empty.

Boston

The Blenkers' house was a big old place. All the windows were open, but it was completely silent. Everyone had gone to the party. As Newland walked through the garden, he saw something pink. Someone had left a pink parasol [1] on the wall. Newland felt absolutely certain that it was Ellen's. He picked up the parasol and put its handle to his lips. Just then he heard the sound of someone approaching: a woman in a rustling silk dress. He didn't look up. He had always known that this might happen...

"Oh! Mr Archer!" cried a loud young voice. Looking up, he saw the youngest and largest of the Blenker girls standing before him. "Where did you come from? No one's home except for me. They all went to the party. Mother said I couldn't go because I have a cold. I was very disappointed, but it's not so bad now that you're here." She smiled at him.

"Has Countess Olenska gone to the party too?" he asked.

"No. She received a telegram yesterday and had to go to Boston." Then she saw the parasol in his hands and cried, "Oh! You've found it! Thank goodness! I've been looking for it everywhere." She took the parasol from him, opened it, and put it over her large blonde head.

"Do you know where Countess Olenska is staying in Boston?" said Newland. "I'm going there tomorrow on business, and I'd like to —"

"How kind of you! She's staying at the Parker House [2]."

When he got home, Newland saw a letter from the office waiting for him on the table by the door. He opened it as he went into the drawing room, where he could hear voices. May and Mrs

1. **parasol** : an object like an umbrella that provides shade from the sun.
2. **the Parker House** : a famous old hotel in Boston. It is still open today.

Welland were back from the party. The letter contained nothing important. When he'd read it, Newland put it in his pocket and said to May, "I've had a letter from the office. They want me to go to Boston tomorrow on business."

The next morning he took the train to Boston and a taxi straight to the Parker House, but the receptionist told him that the Countess was out.

"Out?" repeated Newland, as if it were a word in a foreign language. He left the hotel and went for a walk in the park. As he was walking there, feeling anxious and frustrated, he suddenly saw her sitting on a bench under a tree. She looked rather tired and sad. She was holding a grey silk parasol. How could he ever have thought she'd have a pink one? He walked up to her.

She looked up, startled [1], and said, "Oh!" But then a lovely smile spread over her face. "Oh," she said again, in a different tone.

Newland sat beside her on the bench. "I'm here on business," he said. "What a surprise to see you here!" He didn't know what he was saying. He felt as though he were shouting at her across a large distance and that she might vanish [2] before he could get to her.

"I'm here on business too," she said.

"What business?" he asked.

"Very unconventional business," she said with a smile. "I've just refused to take back a sum of money that belonged to me."

"Your husband has come here to meet you?"

"No! At this time of year he is always at Baden-Baden. He sent

1. **startled** : surprised, for example, by a sudden movement.
2. **vanish** : disappear suddenly.

a messenger. His secretary." She said the word as casually as if it were any other word in her vocabulary. "But I've refused, and I'll go back to Portsmouth by the afternoon train." She looked at Newland for a while then said, "You haven't changed."

He felt like saying, "I had, till I saw you again." Instead he stood up and said, "Let's go out to lunch together. Why not? Haven't we done all we could?"

"You mustn't say things like that to me."

"I'll say anything you like or nothing. I won't open my mouth unless you tell me to. I just want to listen to your voice. It's a hundred years since we met. It may be another hundred before we meet again."

"Why didn't you come down to meet me by the sea that day at Granny's?" she asked suddenly.

"Because you didn't look round — you didn't know I was there. I told myself that I wouldn't go to you unless you looked round." He laughed at his own childishness.

"But I didn't look round deliberately[1]. I knew you were there. When you drove in, I recognized the carriage, so I went down to the beach."

"To get as far away from me as you could?"

"Yes."

He laughed again. "Well, you see, it's no use. My 'business' in Boston was simply to find you. Come on. Let's go to lunch."

At the restaurant, they talked and were silent. The silences weren't embarrassing: they were just as natural as the conversation. She told him what she'd been doing in the eighteen months since they'd last met.

"I was so glad to come home to my friends and relatives in New York," she said, "but, after a while, I realized that I was too

1. **deliberately** : done in a way that was planned.

different to feel at home there, so I moved to Washington. I'll probably stay in Washington. You meet a greater variety of people and opinions there. People in New York blindly follow tradition, and the tradition they follow is somebody else's. Do you think Christopher Columbus would have taken all that trouble to cross the Atlantic if he had known that people in America would make a bad copy of European society?"

She smiled, but Newland felt irritated by her criticism of New York. "Do you say that kind of thing to Beaufort?" he asked.

"I haven't seen him for a long time, but I used to, and he understood me."

"You don't like us," cried Newland, "and you like Beaufort because he's European. You think we're boring. Why don't you go back to Europe?"

He thought she'd be angry with him for saying that, but instead she sat in thoughtful silence for a while and then said, "I stay here because of you."

He blushed and waited in silence, hoping that she would say more.

"At least," she added after a while, "it was you who taught me that under the conventionality there are fine values — that people here care about their families and look after each other in a way that would seem strange where I come from. All the exquisite [1] pleasures of Europe seemed empty and cheap [2] then."

He wanted to say, "At least you've experienced exquisite pleasures! I never have!" But he looked at her in silence.

"I've wanted to have this conversation for a long time," she said. "I wanted to tell you how much you've changed me."

1. **exquisite** : delicate and sensitive.
2. **cheap** : of little value.

"You've changed me too!" cried Newland. "Don't forget: I'm the man who married one woman because another one told him to."

She blushed and said, "You promised not to say things like that to me."

"Ah! How like a woman!" said Newland. "None of you has the courage to talk about the bad things!"

"Is it a bad thing — for May?" she asked.

He heard the tenderness with which she spoke her cousin's name.

"Well," she continued, "didn't you tell me that we always have to think of the feelings of others? We always have to think of the family — to try to make them happy?"

"If you think that my marriage is a success, you're mistaken. If you think that by giving me up you've made May happy, you're wrong! You gave me my first glimpse [1] of real life, and then you told me to continue the false life. No one could endure that!"

"I'm enduring it!" she cried, her eyes full of tears. Suddenly her entire soul — everything she was feeling — was expressed in her face.

"You too? Oh, all this time, you've been going through this misery too?"

For answer, the tears flowed down her face.

"Don't go back to Europe. Please, don't go," he said.

"I won't," she replied, "as long as we can stand it [2]."

He sat in silence, trying to fix her words in his memory. He knew that he would never again feel entirely alone.

1. **glimpse** : brief look.

2. **stand it** : accept or deal with it.

The text and **beyond**

FCE 1 Comprehension check

For questions 1-5, choose the correct answer — A, B, C or D.

1 At the beginning of the chapter we can see that Newland
- A ☐ has almost forgotten about his love for Ellen.
- B ☐ does everything he can to see Ellen.
- C ☐ does not realize how much he desires to see Ellen.
- D ☐ does everything he can to stay away from Ellen.

2 In the sentence "He had always known that this might happen", "this" refers to
- A ☐ a visit to a place where Ellen was staying.
- B ☐ a meeting with a friend of Ellen's who could tell him all about her.
- C ☐ a private meeting with Ellen.
- D ☐ a private meeting with a beautiful young woman.

3 It is significant that Ellen said the word "secretary" so casually because
- A ☐ the secretary worked for her horrible husband.
- B ☐ people had said that she had lived with the secretary.
- C ☐ people had said that the secretary had helped her.
- D ☐ nobody in New York society would have a personal secretary.

4 Thanks to Newland, Ellen now knows that in New York society
- A ☐ there are some very intelligent and artistic people.
- B ☐ tradition is not as important as it seems to outsiders.
- C ☐ people really care for each other.
- D ☐ tradition helps keep family and friends together.

5 When Ellen says that she will not return to Europe "as long as we can stand it", the "it" refers to her
- A ☐ staying near Newland without having a love affair with him.
- B ☐ living in cities without the great cultural traditions of Europe.
- C ☐ being nice to people that she does not respect.
- D ☐ living without her husband's money.

2 False life versus real life

What is the real life and what is the false life for Newland? For Ellen? Discuss your opinions with your friends.

A Passionate love with a person who shares his/her interests in life.

B Respect for those who love him/her and trust him/her.

C Life with moments of real exquisite happiness and pleasure, even if he/she must lie and deceive.

D The thought and occasional presence of the person he/she really love.

E *Your own idea ...*

FCE 3 Sentence transformation

For questions 1-10, complete the second sentence so that it has a similar meaning to the first sentence, using the word given. Do not change the word given. You must use between two and five words, including the word given. There is an example at the beginning (0).

0 But you're the woman I would have married if it had been possible for either of us.
WAS
I didn't marry you because it was not possible for either of us.

1 All I feared was bringing scandal on the family.
ONLY
The .... afraid of was bringing scandal on the family.

2 I didn't know that they all refused Granny's invitation to a dinner to meet me.
DOWN
I didn't know that they .... Granny's invitation to a dinner to meet me.

3 She told me that you and May announced your engagement early so that I would have the protection of two families, not just one.
ME
She told me that you and May announced your engagement early so that two families .... not just one.

4 But no one was as kind as you.
WERE
But .... of all.

5 "I won't be lonely anymore," she said.
SHE
She said that .... lonely anymore.

6 Many husbands find it impossible to join their wives here except at the weekends.
CAN
Many husbands .... their wives at the weekends.

7 I've never heard you say anything cruel before.
EVER
This is the .... heard you say anything cruel.

8 Mrs Archer said they'd all been invited by Professor and Mrs Emerson Sillerton to a party for the Blenkers.
BEEN
"We .... by Professor and Mrs Emerson Sillerton to a party for the Blenkers," said Mrs Archer.

9 The rest of the afternoon was free.
NOTHING
He .... the rest of the afternoon.

10 No one's home except for me.
ONLY
I .... home.

4 Summary

Fill in the gaps with a phrase to make true sentences and to complete the summary of chapters Four to Six.

At his office, Mr Letterblair asked Newland to talk with Countess Olenska about her divorce. He was supposed to explain to her that her (0) family did not want a scandal . Letterblair then gave Newland all the letters and papers of the case.
The next day, Newland went to talk with Ellen. When he arrived, he was disappointed to find her with Julius Beaufort. Soon, though, Newland and Ellen were alone. They talked for a moment about art, but then Newland mentioned the divorce. At first Ellen was very happy: she thought that (1) .... But then he explained her family's point of view.
After this, Newland's feelings for Ellen grew stronger. He even went down

to St Augustine in Florida to ask May to (2) ....
This request made May think that he (3) ....
When Newland returned to New York City, he went to see Ellen. He told her what May had said. She asked if there was another woman.
Newland then admitted that he loved Ellen, and that he wanted to marry her and not May. Ellen refused, even though she loved him too, because (4) ....
Just then Ellen received a telegram from May. It said that (5) ....
....
So, Newland and May were married. They went on their honeymoon and returned to their normal married life in New York. Gradually, Newland's love for Ellen began to seem like (6) .... A year after his wedding in Newport, he was even able to resist going down to the seaside to see Ellen. However, when Newland discovered that Ellen had gone to Boston, he made an excuse to go there to see her.
When he arrived in Boston, by chance he found her sitting on a park bench. She looked sad and tired. They decided to go out for lunch. It was then that Newland discovered that she too (7) .... She told Newland that she would stay in America as long as they didn't become lovers.

Before you read

1 Listening

You will hear a conversation between Newland and Sillerton Jackson about Julius Beaufort going bankrupt. For questions 1-8, complete the sentences.

May, Newland and Sillerton have dinner at (1) ....
Newland and Sillerton begin to talk after (2) ....
Newland says that it is no secret that (3) ....
Sillerton thinks that Ellen should have (4) ....
Newland becomes angry but Sillerton is quite (5) .... as he smokes.
Sillerton thinks that now Ellen does not have enough (6) ....
....
Ellen invested her money with (7) ....
Newland is worried that he has revealed his true feelings for Ellen, and so he suggests (8) ....

CHAPTER **SEVEN**

Beaufort's Disgrace

Four months passed, and Newland didn't see Ellen again. She went back to Washington, and he went back to his false, empty life. One evening, he and May went to dinner at his mother's house with Sillerton Jackson. When the ladies had left them to their brandy and cigars, Mr Jackson said, "It looks as though Beaufort will go bankrupt. If that happens, it will be a big scandal. He didn't spend all that money on Regina."

"Well, everyone knows that," replied Newland impatiently.

"It's a pity that Countess Olenska didn't accept her husband's offer."

"Why do you think it's a pity?"

"Well, what's she going to live on now? If Beaufort — "

Newland leapt to his feet [1] and banged his fist on the table. "What the devil do you mean, sir?" he cried indignantly [2].

1. **leapt to his feet** : stood up quickly.
2. **indignantly** : expressing anger and surprise because you feel insulted or unfairly.

Mr Jackson smoked his cigar and looked serenely at the young man's angry face. At length he said with a smile, "Well, she hasn't got much money, and what she did have was invested with Beaufort. So I could ask you, my dear boy, what do you mean by asking me what I mean?"

"You know perfectly well that what you said seemed to suggest — "

"Yes, but I'm not the only one who's suggesting it. Larry Lefferts told me, and he isn't the only one talking about them either."

Newland was afraid of having shown too much to this observant [1] old man. "I think it's time we joined the ladies," he said.

END

*

Newland decided to go to Washington to visit Ellen. He could wait no longer. He had to see her. He told May that he had business in Washington and would be gone for several days. He made arrangements to leave on Tuesday.

Early on Sunday morning, however, Mrs Manson Mingott had a stroke [2]. On Saturday evening, Regina Beaufort had come to visit her. She'd begged Mrs Mingott to lend Beaufort the money he needed to avoid bankruptcy. She'd insisted that the whole family's honor depended on this. "I'm a Dallas!" Regina had cried to her aunt.

"No, Regina!" the old lady had replied. "Your husband has ruined himself and hundreds of innocent people who trusted him

1. **observant** : good or quick at noticing things.
2. **a stroke** : an illness caused by a blocking of the blood flow in your brain.

with their money. He has brought shame [1] on anyone associated with him. You were a Beaufort when he covered you in diamonds, and you're still a Beaufort now that he has covered you in shame!"

At three in the morning, Mrs Mingott had called her maid. The maid found her sitting up in bed, unable to speak properly or to move her left arm. She sent a messenger to the doctor's house and to the houses of Mrs Welland and Mr Lovell Mingott — the old lady's children. The rest of the family arrived at six and went in to see her one by one. They were relieved to see that she was a little better. She could speak clearly now. When Mr Lovell Mingott came out of his mother's room, he said, "She said she'll never speak to Regina Beaufort again. She also says we must send a telegram to Ellen, telling her to come here immediately." The whole family was rather shocked by this last bit of news. They felt offended that their own presence wasn't comfort enough. Obviously the old lady cared for no one but Ellen. They were also alarmed: clearly if the old lady asked for Ellen she must be very ill indeed. She must be afraid of dying. Why else would she command Ellen to come to New York? May thought perhaps her grandmother wanted to try one last time to persuade Ellen to return to her husband.

Newland listened to these discussions in silence. "Will you go to the Post Office, Newland, and send the telegram?" asked May.

"Of course," Newland replied.

As he walked to the Post Office, Newland saw Beaufort's disgrace announced on every newsstand. The whole of New York

1. **shame** : the opposite of "pride"; the feelings of embarrassment and self-disgust you get when you know that everyone disapproves of you.

was shocked at his dishonor, and fashionable young gentlemen were gossiping about it on every corner.

The next day, a telegram arrived in reply to the one Newland had sent. It said that Countess Olenska would arrive at Jersey City station on Tuesday evening.

"Somebody must meet her. It's two hours' drive from Jersey City. We can't let her come back to New York on her own," said Mrs Welland. "Lovell and I must be here with mother, and Mr Welland isn't well enough to go."

"I'll go," said Newland.

"But Newland!" cried May. "You'll be in Washington then. You told me you were leaving on Tuesday morning for a business meeting."

"It's been cancelled," said Newland. "So, I can go and meet Countess Olenska. It's no trouble at all."

"Really?" said May in amazement. "What a coincidence!" She looked at him, and her eyes at that moment seemed so blue that Newland wondered if there were tears in them.

"Oh, thank you so much, Newland," said Mrs Welland, and May looked down.

The text and **beyond**

1 Comprehension check

Answer the questions below.

1 What did Mr Jackson mean when he said, "He didn't spend all that money on Regina"?
2 Why did Newland become angry with Mr Jackson?
3 How did Mr Jackson defend himself?
4 How did Newland justify his trip to Washington?
5 What caused Mrs Mingott's stroke?
6 Why, according to Regina, should Mrs Mingott have helped Beaufort?
7 How did Mrs Mingott respond?
8 Who did Mrs Mingott send for once she was feeling better?
9 How did May react when Newland offered to meet Ellen?

2 The Panic of 1873

The character Julius Beaufort, with his European charm and his love of women and horses, was based on the financier August Belmont (1813-1890). However, Beaufort's disgrace and the financial troubles it caused were inspired by the economic disaster known as "The Panic of 1873".
It was in this period that the old wealthy families of New York had to worry that a financial crime by some Beaufort could lead them to poverty. This is the new, nervous world described in *The Age of Innocence*.
Read the text about the Panic of 1873, and fill in the gaps with the words in the box. There are five words that you do not need to use.

also	blame	becoming	even	acted	owned	amount
wrong	lasted	manage	deals	occurred	increase	reached
more	other	came	became	another	between	involved

The economy of the northern states after the American Civil War (1861-1865) was extremely strong. To defeat the Confederacy (the southern states) the Union (the northern states) had to (1) .... its industrial production. Then after the war, northern industry had (2) .... great stimulus: the railroads. (**3**) .... 1866 and 1873 about 35,000 miles of tracks were laid. In 1869 the first transcontinental railroad was completed. What is more, railroads had become the second largest employer in the United States after agriculture.

But all this immense economic activity was not done with gold and silver, or (**4**) .... with individual banknotes. It was done with credit, or the promise to pay or work. As one journalist wrote in 1873, "In the ancient world, what a man (**5**) ...., he held. His wealth was his farm, or his house, or his ships. He could hardly separate the idea of property from that of possession." "But now," he continued, "The wealth of a man are some papers he has signed declaring that he has a right to so many shares of a distant bank, or that some railroad will pay him a certain (**6**) .... of money in thirty years." This system of credit and shares in companies was the only way to (**7**) .... a large economy, but it could be manipulated at times in very dishonest ways. A master of this dishonest manipulation was Jay Gould (1836-1892). One of Gould's most spectacular (**8**) .... was when he tried to corner the gold market[1] in 1869. His plan just barely failed. The economic disaster that followed (**9**) .... on September 24, 1869, and is known in history as *Black Friday*.

Gould was also (**10**) .... in the Panic of 1873. His banking firm, Gould and Company, went bankrupt when it tried to finance another transcontinental railroad. After his firm's collapse, 37 other banks closed, and the New York Stock Exchange was closed for five days. Then 89 of the nation's 365 railroads collapsed. This was the beginning of the *Long Depression* which (**11**) .... for five years. During this time about 18,000 businesses went under and unemployment (**12**) .... 14 percent. Jay Gould was not, of course, to (**13**) .... for all this. But it was (**14**) .... clear in the 1870s that the huge and powerful modern economy was also a very delicate thing, often based on nothing (**15**) .... than people's faith that other people would do their job well and that behind pieces of paper there was real money. But when faith went, panic (**16**) ....

1. **he tried to corner the gold market** : he bought a large quantity of gold in order to control the market.

3 **Bank Runs**

One of the first signs of the arrival of a major financial crisis was a bank run, like thunder before a rain storm. A bank run occurred when the customers of a bank were afraid that their bank did not have enough money to pay its debts. Even when a bank had no financial problems, it never had enough money to give all its customers all their deposits at the same time. So, if all the customers came running and asking for their money, the bank would have to close. The closing of one bank then caused the closing of another bank, and the crisis began. The phenomena of bank runs and panics were great disasters feared by everybody, both rich and poor.

Of course, the producers of plays exploited this fear to sell tickets to their plays, such as the 1895 play *The War of Wealth.*

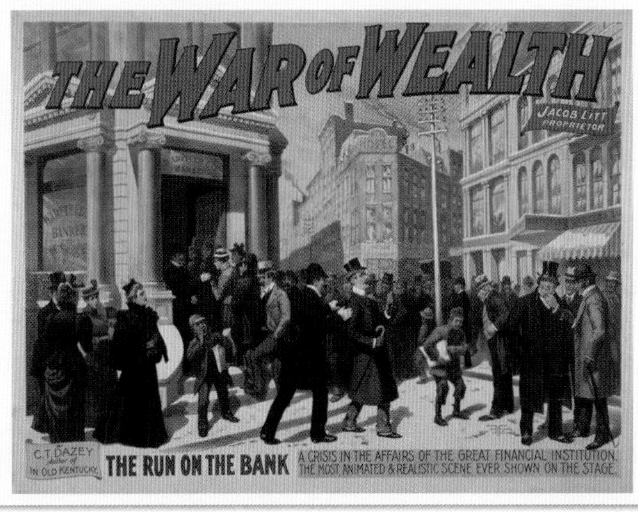

With your partner, make a short presentation about the poster above. Use the questions below to help you.

1 What are the two boys selling?
2 What do you think the policeman (the man with the white gloves) is doing?
3 How do the three men in the bottom right-hand corner look?
4 What is special about how they show a crisis in a 'great financial institution'?
5 What real-life disasters have been used in films?
6 Do you like this kind of film? Why or why not?

Before you read

1 Prediction

In Chapter Eight Newland meets Ellen at the station. When they're in the carriage, he says, "*... each time I see you, you happen to me all over again.*" What do you think he means by these words?

2 Listening

7 FCE

You will hear a conversation between Ellen and Newland. For questions 1-6, choose the correct answer — A, B or C.

1 When Ellen heard about Beaufort's bankruptcy she was sorry for
- A ☐ him.
- B ☐ herself.
- C ☐ his wife.

2 When Newland explained what he meant by "I hardly remembered you", Ellen
- A ☐ didn't agree.
- B ☐ understood him.
- C ☐ didn't understand.

3 Newland finally held Ellen in his arms because
- A ☐ the movement of the carriage pushed her.
- B ☐ she asked him to.
- C ☐ she seemed cold to him.

4 Newland suggested getting out of the carriage because
- A ☐ he wanted more time with Ellen.
- B ☐ they were too obvious in the carriage.
- C ☐ he realized that she was uncomfortable in May's carriage.

5 Ellen refused to walk with him because
- A ☐ it was too cold out and she was not dressed properly.
- B ☐ she was afraid to be seen by someone they knew.
- C ☐ the purpose of her visit was to see her grandmother.

6 Ellen shocked Newland because
- A ☐ she used the word "mistress".
- B ☐ she refused to be his mistress.
- C ☐ she offered to be his mistress.

CHAPTER **EIGHT**

Snow

It was a dark, snowy afternoon. When Ellen got off the train, Newland was startled by her pale face. She looked at him in surprise. "Come," he said. "I have a carriage waiting."

He hurried her into the carriage and sat down beside her. He told her all about her grandmother's illness and Beaufort's ruin. "Poor Regina!" she said softly.

"Were you surprised to see me at the station?" he asked then.

"Yes!" she replied with a smile.

"So was I," said Newland. "When I saw you, I was surprised. I hardly remembered you."

"Hardly remembered me?" she repeated in amazement.

"I mean — how can I say it? — each time I see you, you happen to me all over again."

"Oh, yes. I know!" she said.

"Is it the same for you?" he asked.

"Yes," she said, looking out of the window.

"Ellen, Ellen, Ellen!"

She didn't reply, so he sat in silence, watching her profile against the snowy window. The precious moments were slipping away, and he'd forgotten everything he'd planned to say to her.

"What a pretty carriage!" she said after a while. "Is it May's?"

"Yes."

"Did May send you to meet me? How kind of her!"

With a jolt [1], the carriage went over a bump in the road, and she was thrown against him. He put his arm around her and said, "We can't go on like this, Ellen. We can't be together and not be together."

"You shouldn't have come today!" she cried. Suddenly, she threw her arms around his neck and kissed him; then she turned away and looked out of the window, trying to keep as far away from him as possible.

"Don't be afraid of me," he said. "I know you don't want a squalid [2] affair, and neither do I. I want us to be together — really together — not just for an hour in secret with days of longing in between."

"You chose a good place to tell me that!" she said, laughing.

"Why? Because this is my wife's carriage? All right. Let's get out and walk. I'm sure you're not afraid of a little snow."

"No. I won't get out and walk, because I must go to Granny's. That's what I'm here for. And you'll sit beside me and talk to me, not about visions but about realities."

"This is the only reality," he said.

She sat silent as the carriage turned into Fifth Avenue. "Do you want me to live with you as your mistress, since I can't be your wife?" she asked.

1. **jolt** : sudden sharp movement.
2. **squalid** : involving low moral standards or dishonesty.

The crudeness of the question startled him. Women in New York high society never used that word, even when they were talking about Fanny Ring. Ellen spoke it clearly and simply as if it were a normal word in her vocabulary.

END

"I want to go away with you to a world where words like that — ideas like that — don't exist, a world where we'll just be two human beings who love each other."

She laughed again. "Oh, my dear — where is that country?" she said. "Have you ever been there?"

The carriage had crossed Forty-Second Street. May's excellent horses were pulling them quickly to the end of their journey. Soon the precious two hours would be over. "So what is your plan for us?" asked Newland.

"For us? There is no 'us', in that sense. We're near each other only if we stay far from each other. Then we can be ourselves. Otherwise, we're only Newland Archer, the husband of Ellen Olenska's cousin, and Ellen Olenska, the cousin of Newland Archer's wife, trying to be happy behind the backs of [1] the people who trusted them."

"I'm beyond that," said Newland.

"No you're not!" cried Ellen. "You've never been beyond. But I have! And it isn't a place you and I want to be."

He sat in silence, full of pain. Then he called to the driver and asked him to stop the carriage.

"Why are we stopping? This isn't Granny's!" said Ellen in surprise.

"No, but I'll get out here," he replied, opening the door and

1. **behind the backs of** : deceiving; to do something behind people's backs is to do it without telling them, in secret.

stepping out into the snow. "You're right. I shouldn't have come to meet you," he whispered to her, then he called to the driver, "Drive on!"

*

That evening, May returned from Mrs Mingott's house just before dinner. Newland and May dined alone. "Why didn't you come back to Granny's?" asked May, as the servant filled her wine glass.

"I had some letters to write. Besides, I didn't know you were staying there. I thought you'd be at home."

She didn't reply, and he noticed that she looked tired and sad. For the first time, Newland thought that perhaps the monotony [1] of their life together caused her pain too. Then he remembered that, as he was leaving Mrs Mingott's house to go to Jersey City, she'd said to him. "I'll see you back here, then." He'd replied, "Yes," but he'd immediately forgotten about it. He had had other things to think of. Now he felt slightly irritated that she should be offended by so trivial an omission after two years of marriage.

After dinner they went to the library for coffee. He sat down to read a history book. When they were engaged, he used to read poetry aloud to her, but after their marriage he'd stopped doing that: her comments on the poems were too depressing. Now he preferred to read history in peace. May took out her embroidery [2] and started working on it. She wasn't very good at embroidering, but all the other wives embroidered cushion-covers for their husbands, so May did it too. Every time he looked up from his book, there she was. Her sapphire engagement ring and gold wedding ring gleamed in the lamp light. "She'll always be the

1. **monotony** : a lack of variety that makes you feel bored.
2. **embroidery** : the activity of stitching designs onto cloth. (See picture on page 91).

same", thought Newland. "In all the years to come, she'll never surprise me with a new idea or emotion." She was maturing into a copy of her mother. He put down his book, went to the window, opened it, and leaned out into the icy air. The snow was still falling; big soft flakes of it were blowing in the wind.

"Newland!" said May. "What are you doing?"

"I want some fresh air. It's stifling [1] in here."

"Please shut the window. You'll catch your death!" [2]

He shut the window and turned to her. "Catch my death!" he repeated in a sarcastic [3] voice. He wanted to say, "I've caught my death already! I've been dead for months and months!" But then he had another thought: "What if May dies? Young healthy people like May sometimes get ill and die. What if that happened to May? Then I would be free!"

She looked up at him in surprise. "Newland, are you all right? Are you feeling ill?"

"No," he said, returning to his chair, and as he passed by her he put his hand on her hair. "Poor May!"

"Poor? Why poor?" she said with a nervous laugh.

"Because I'll never be able to open a window without worrying you," he replied.

For a moment she was silent and then she said, "I'll never worry if you're happy."

"Ah, my dear! I'll never be happy unless I can open windows!"

"In this weather?" she cried.

Newland sighed and returned to his book.

1. **stifling** : so hot and airless that you can hardly breathe.
2. **You'll catch your death!** : "You'll catch your death of cold" means you will catch a very bad cold.
3. **sarcastic** : saying things that are the opposite of what you mean in order to be unkind to sb or to make fun of them.

The text and **beyond**

1 Comprehension check

Who said what and why? Match the quotes with the character who said them, and then match the quotes with the reason why they said them. You may match some characters with more than one quote. There is an example at the beginning.

Who

Newland Archer (N)   May Archer (M)   Ellen Olenska (E)

What

A [N] [6] "I hardly remembered you."
B [ ] [ ] "How kind of her!"
C [ ] [ ] "I'll never worry if you're happy!"
D [ ] [ ] "Oh, my dear — where is that country?"
E [ ] [ ] "I caught my death already!"
F [ ] [ ] "I had some letters to write."
G [ ] [ ] "You chose a good place to tell me that!"
H [ ] [ ] "I'm beyond that."

Why

1 He/She is trying to discreetly remind Newland of their responsibilities to other people.
2 He/She is hiding his/her meeting with his/her lover.
3 He/She is saying that he/she doesn't care about the feelings of his family.
4 He/She is saying that it is impossible for people to live without worrying about family, friends and society in general.
5 He/She is saying that what they are doing at the moment is squalid.
6 He/She is saying that seeing the other person has not become a habit and is always surprising and new.
7 He/She is saying that he/she cannot stand the closed world of New York society and his/her marriage.
8 She is trying to bring Newland back into her life.

2 Word puzzle

Complete the words in the categories below, and then find these words in the word square.

Being surprised

s _ oc _ _ _
_ l _ r _ e _
_ _ _ _ tl _ _

Love outside marriage

_ ff _ i _
mis _ _ _ _ _

Illness

_ tro _ _
_ ol _

The sea

_ _ gh _ h _ _ s _
_ a _ l _ _ _ t
_ ie _

Marriage

c _ r _ m _ n _
e _ g _ g _ _
_ a r r _
_ _ _ _ y _ oo _

Scandal

_ isg _ _ _ _
_ _ _ me
d _ sh _ _ _ _
_ iv _ r _ _ _

Transportation

car _ _ _ _ _
_ tat _ _ n
r _ _ lw _ _

```
M A R I P A F F A I R T U N N Y F H O R
D D I S H O N O R G L I M P S E M O I T
S I S H U P C A R R I A G E O M I N M H
F H S U N D I E R W I T H E G M S E G E
E I J C C E R E M O N Y C R M O T Y R B
S L G M O N F O R E G E T T A N R M A S
A M E U B G G I K L S L I L K R E O I T
I S E P R A T O S U C M I B I Y S O L A
L S H T H G Z O L O C A L G A I S N W R
B T T O C E C O L L D B U S N S R A A T
O R S R C D L I G H T H O U S E M A Y L
A E T B O K E L O M Y H I A D B A C H E
T S R D B K E B M Z A X O P C E P A S D
P S O B I V E D Q U E R O P E O L D S G
E S K R B V A L O S E D R E S T L L S I
A L A R M E D M O A N I S Y E T O D T S
R L M S T A T I O N M A N N I E S O A L
S D I S G R A C E B S H A M E R B V R I
A B C A T I N N G O N T T H A T O E V S
L B D I V O R C E D H O N E I H M D B H
```

T: GRADE 8

3 Speaking: public figures

When Edith Wharton was young, the wealthy people of New York knew about everybody else's personal problems. It was, after all, a very small community. But their private business was also known, in part, to the general public. Their marriages, divorces and love affairs and expensive entertainments, could easily become public through the newspapers. Today, the life of the rich and famous is a favorite with the general public. Give a short report about the media's interest in celebrities in your country. Use the following questions to help you.

1 Are there special magazines and newspapers that talk mostly about celebrities?
2 Do regular news, shows and newspapers also dedicate much space to these things?
3 What is a big scandal now?
4 Why do you think people are so interested in the private lives of famous people?
5 Do these things interest you? Why or why not?

Before you read

1 Listening

Look at the six statements below. Then listen to the beginning of Chapter Nine. You will hear about Newland's visit to Mrs Mingott. Say whether the statements are true (T) or false (F).

		T	F
1	Newland is upset that nobody in the family talks about Ellen.	☐	☐
2	Newland is surprised when Mrs Mingott wants to see him.	☐	☐
3	Newland feels excited before he enters Mrs Mingott's house.	☐	☐
4	Mrs Mingott jokes about Newland complimenting Ellen.	☐	☐
5	The family is happy that Ellen will live with her grandmother.	☐	☐
6	Mrs Mingott sees Ellen as a little bird that needs to be free.	☐	☐

CHAPTER **NINE**

Decisions

Six or seven days passed. Newland didn't see Ellen and no one in the family mentioned her name in his presence. He didn't mind. He could wait. That night when he'd leaned out of the window in the icy cold, he'd made a decision. When Ellen left New York, he would leave with her. He would go with her to Washington or somewhere else, if she agreed. Japan, for instance. They could go to Japan. For days he'd been thinking about this plan. Then one day May told him that Mrs Mingott wanted to see him. There was nothing strange about the request: she was getting better, and Newland was her favorite grandson-in-law. It was natural that she would ask to see him.

Standing outside Mrs Mingott's door, Newland felt his heart beating fast. In another minute he would see Ellen. He would speak to her. He would ask her when she was going back to Washington.

The maid answered the door and took him into Mrs Mingott's room. He looked around for Ellen, but she was nowhere to be seen. The old woman sat in an enormous armchair by her bed. She was pale, and there were dark shadows under her eyes, but she was much better than she'd been that first morning after the stroke. She cried out in delight when she saw him, "Hello, my dear! Am I terribly ugly?"

"You're prettier than ever!" replied Newland, laughing and taking her hand.

She laughed too and said, "But not as pretty as Ellen! That day when she came from Jersey City, she looked very pretty. I thought perhaps you'd told her so, and that's why she made you walk home in the snow."

She was still laughing, so he laughed too, waiting for the joke to be over.

"It's a pity she didn't marry you!" said Mrs Mingott suddenly. "Then I wouldn't have had all this worry about Olenski."

Newland wondered if her illness had affected her brain.

"Anyway, it's all over now. She has agreed to stay here with me. You know, they all tried to persuade me to cut off her allowance [1], so that she would have to go back to her husband. Yes, they did! Lovell and Augusta and Letterblair and the rest — they all tried to persuade me. And they nearly did persuade me, especially after that secretary came with the Count's new offer. It was a very generous offer, and I thought, 'Money is money, and marriage is marriage!' But when I saw her, I thought, 'You sweet bird, we can't put you back in that cage again!' I decided

1. **allowance** : money that is regularly given to someone in order to help them pay for the things they need.

then that I wouldn't force her to go back. Now she has agreed to stay and take care of her Granny, and of course I've told Letterblair that she must have her proper allowance."

END

As Newland listened, his heart beat fast. At first, he felt confused and perplexed [1]. He'd made a decision, and now everything was changed. Then slowly he realized that this change made things easier. If Ellen had agreed to come and live with her grandmother, surely it meant that she now understood that they couldn't live apart anymore. This was her answer to what he'd said to her in the carriage. She wouldn't take the extreme step of running away to live with him, but she would come back to New York so that they could see each other more frequently. He'd been ready to risk everything to be with her, but that was no longer necessary.

"The family are still opposed to it," Mrs Mingott continued. "They still want her to go back to her husband, and they say that I'm too old and too ill to make a proper decision about it. You'll have to help me, Newland."

"Why me?" asked Newland.

"Why not?" The old lady looked at him with her quick intelligent eyes.

"I'm too insignificant. They won't listen to me."

"You're Letterblair's partner. You must persuade Letterblair to persuade them!"

"I'll try my best."

"Good. I knew you would, because they never quoted you when they were saying that everyone thought it was her duty to go home."

1. **perplexed** : confused and worried by sth you do not understand.

He wondered if they had quoted May, but he didn't ask.

"Is Countess Olenska in?" he asked instead.

"No. She went to see Regina Beaufort. I told her that I'll never speak to Regina Beaufort again, but she said, 'Come on, Granny. She's your niece, and she's a very unhappy woman.' Then I said, 'And she's the wife of a very bad man!' and Ellen replied, 'So am I, and my family want me to go back to him!' Well, I didn't know what to say to that, so I lent her my carriage and let her go."

"I have to go now," said Newland. He kissed the old lady's hand, which was still in his.

"Ah! Whose hand do you imagine you're kissing?" cried Mrs Mingott, laughing at him. "Your wife's, I hope!"

*

He left Mrs Mingott's house and walked quickly to the Beauforts' on Fifth Avenue. He remembered the house blazing with lights on the night of the Beauforts' ball, when he and May had announced their engagement. Now it looked dead. There was only one lighted window. Some people were saying that Beaufort had left New York with Fanny Ring, but that seemed improbable. Mrs Mingott's carriage was waiting outside the door. Newland felt full of admiration for Ellen: she alone had rushed to Regina's side to show her solidarity and affection for her cousin in troubled times.

Suddenly the front door opened, and she came out. She turned and said something to someone inside the house, then she descended [1] the steps.

"Ellen," he said in a low voice.

1. **descended** : moved downwards.

She stopped in surprise. He noticed two men walking along the other side of the street. They passed under a street lamp, and he saw they were Larry Lefferts and Sillerton Jackson. They looked over at Newland and Ellen with interest.

"Tomorrow I must see you somewhere where we can be alone," he whispered.

She laughed. "In New York?"

"There's the Art Museum in the park," he said. "I'll wait for you at the door at half past two."

She turned away without answering and got into the carriage.

*

The next day they met at the museum. "Why have you come back to New York?" he asked.

"Because I thought at Granny's house I would be safer," she said.

"Safer from me?"

She looked down at her hands and didn't reply.

"Safer from loving me?" he asked.

He saw that her eyes were full of tears. "Safer from doing irreparable harm!" she cried. "Let's not be like the others!"

She meant all the people in New York society who had affairs and lived a life of lies with their husbands and wives. She didn't want to be like those people, and neither did he, but something made him say, "What others? I'm no different from anyone else. I have the same desires."

She blushed and looked at him. "Shall I come to you once, then go home?" she asked in a clear low voice.

14

"Oh my dearest!" he said, but then he hesitated. "What do you mean by 'go home'?"

"Home to my husband."

"No! Of course you can't go home!"

"Well, I can't stay here and lie to people who have been good to me."

Newland looked at her in despair. It would be easy to say, "Yes, come once." He was sure that he would be able to persuade her later not to go back to her husband. But he couldn't deceive[1] her. He wanted to be as honest as she was.

"That's why I want you to come away with me," he said. "What we're trying to do is impossible."

She stood up and said, "I must go."

He held her wrist. The thought of losing her was unbearable. "Well, then. Come to me once," he said. For a moment they looked at each other like enemies.

"Send me a note saying where and when," she said.

"Will you come to me tomorrow?"

She hesitated. "The day after," she said finally. Her face was very pale but full of love.

1. **deceive** : trick sb by behaving in a dishonest way.

The text and **beyond**

1 Comprehension check

Match the phrases in column A with the phrases in column B to make true sentences. There are three phrases in column B that you do not need to use.

A

A ☐ Mrs Mingott wanted to see Newland
B ☐ Mrs Mingott knew she could count on Newland's help
C ☐ Newland was very nervous just before he saw Mrs Mingott
D ☐ Mrs Mingott asked Newland about her appearance
E ☐ Mrs Mingott laughed when she talked about Ellen
F ☐ Ellen's family did not want her grandmother to give her money
G ☐ Ellen's grandmother gave her money
H ☐ Mrs Mingott asked Ellen to look after her
I ☐ Newland thought Ellen accepted her grandmother's offer
J ☐ Ellen returned to her grandmother's house

B

1 because she needed his support in her decision to not send Ellen away.
2 because she had just recovered from a stroke.
3 because she could not bear the idea of being far from him.
4 because she could not bear the idea of Ellen going back to Count Olenski.
5 because she knew that Newland loved her.
6 because she was afraid she would cause a scandal in New York.
7 because he thought Ellen might be with her.
8 because he thought she knew about his being in love with Ellen.
9 because they wanted her to leave New York.
10 because she wanted him to help her send Ellen back to her husband.
11 because her family would always be watching her there.
12 because she did not want her to be under her husband's control again.
13 because nobody ever said that he thought Ellen should go home.

2 Edith Wharton remembers

Later in life Edith Wharton wrote about her childhood in New York. Listen to her describe her experience of an elegant mistress, a real-life Fanny Ring. For questions 1-7 complete the sentences.

Edith was (1) .... years old when she saw the yellow carriage.

She saw it on (2) ....

It was difficult for her to see (3) ....

Edith's mother (4) .... when Edith pointed to the lady.

This elegant lady was (5) ....

Edith always did what her mother said and so she (6) .... when this lady appeared in the carriage.

This woman, who Edith saw just once, became like a (7) .... for Edith.

T: GRADE 7

3 Speaking: early memories

Above you have heard Edith Wharton talking about an important early memory that affected her whole life and that also appeared in her novel *The Age of Innocence*. What about you? What incident or person from your childhood has had a big effect on your life? Present your memories to the class. Use the questions below to help you.

- How old were you?
- Where were you?
- Who was this important person?
- What happened?
- How has the event or person affected your life?

Before you read

1 Listening

10 FCE

You will hear about Ellen's future plans. For questions 1-6, choose the correct answer — A, B or C.

1 Newland wanted to leave the Academy of Music because
- A ☐ he didn't like the music.
- B ☐ he didn't feel very well.
- C ☐ he had something to ask May.

2 Ellen's problems had been solved because
- A ☐ she was going to Paris to divorce her husband.
- B ☐ she was going to Europe with enough money.
- C ☐ her husband was giving her money back.

3 May thought Newland would have heard the news about Ellen
- A ☐ at work.
- B ☐ from Sillerton Jackson.
- C ☐ from Ellen herself.

4 May got the news from
- A ☐ Granny, Mrs Mingott.
- B ☐ Ellen herself.
- C ☐ her mother.

5 Newland felt that he could now go to Ellen in Paris because
- A ☐ she wasn't returning to her husband.
- B ☐ nobody in New York would ever know.
- C ☐ May accepted his relationship with Ellen.

6 At the party, Newland thought that Ellen's face
- A ☐ had lost most of its beauty.
- B ☐ had become even more beautiful.
- C ☐ looked strangely different.

CHAPTER **TEN**

A Farewell Dinner

The next evening *Faust* was playing at the New York Academy of Music again. Everything was as it had been two years before on the evening that he and May had announced their engagement: Newland stood in the box rented by his gentlemen's club with Mr Jackson and Lefferts; May was in her grandmother's box with her mother and aunt; the large blonde soprano on the stage sang out "M'ama!" triumphantly. Newland looked at May. He had a strong desire to tell her the truth and ask for the freedom he had once refused, when she offered it to him in St Augustine.

He ran through the red corridor to Mrs Mingott's box. Entering quietly, he leaned close to May and said, "I have a terrible headache. Will you come home with me?"

May whispered to her mother who nodded sympathetically, and in fifteen minutes they were at home, in the library. Newland lit a cigarette.

"I want to tell you something, May."

"Yes, dear?" said May, sitting down.

"It's time I told you something about myself," he began. "Countess Olenska —"

A Farewell Dinner

"Oh, why should we talk about Ellen tonight?" said May impatiently. "I know I've been unfair to her. You've understood her better than most people. You've always been kind to her. But it doesn't matter now, does it, now that it's all over?"

"What do you mean?" asked Newland.

"Well, she's going back to Europe soon, and Granny approves of the idea and has given her enough money to be independent of her husband. I thought you knew. I thought they would have told you at the office." She blushed and looked down.

"How do you know?" Newland asked at last.

"Ellen and I had a long talk yesterday evening, when I was at Granny's house. She was so kind to me. I think she understands everything. Then, this afternoon, she sent me a note. She has gone back to Washington to pack her things. She says she is going to sail from New York for Europe in ten days. She's going to live in Paris. You know, Newland, we haven't given a big dinner party yet. Let's give one for Ellen before she leaves."

*

The dinner was very formal and elegant. Lamplight shone on the ladies' bare shoulders and on their red and blue and gold silk dresses. The gentlemen wore elegant black jackets with white flowers. There were five vases full of orchids on the long dining table. Nine days had passed since Ellen had returned to Washington, and Newland had heard nothing from her. Now she was back in New York, and the next day she would sail for Europe, but she wasn't going back to her husband. Therefore, he could follow her. And, if he did that, he was sure she wouldn't send him away. This thought gave him the strength to get through the dinner party. As the guest of honor, Ellen sat on Newland's right at dinner. Her face looked pale and tired — almost ugly — and he had never loved it so much. He looked down at her hands. All the beauty that seemed to have deserted[1] her face was there in her hands: her long pale fingers and slender[2] wrists. He thought to himself, "I would follow her just for the chance of seeing her hands again!"

END

Now that she was leaving, everyone was affectionate. People kept mentioning her name, referring to her as "Ellen" and "the dear Countess", as though they had never gossiped about her, never thought her life was scandalous, never said she should go back to her husband. Newland looked at them: his eyes went from one well-fed face to another, and suddenly he realized that, for months, all these people had believed that he and Ellen were lovers. And now they were glad because they had succeeded in separating the lovers. It was New York's way of doing things, without scenes[3], without scandal. They had all come this evening to support May, and May understood this. She too believed that Newland and Ellen were lovers. She too approved of this way of avoiding unpleasant scenes. Ellen would go back to Europe, and everything would return to normal.

Mrs van der Luyden sat on Newland's left. She, Mrs Welland and Sillerton Jackson were discussing Beaufort and Regina. They were merciless[4]. "They're doing this," thought Newland, "to show me what would happen to me if I offended them." He laughed out loud.

"Do you think it's funny?" asked Mrs van der Luyden indignantly. "Well, I suppose Regina's idea of staying in New York has its ridiculous side..." and she continued talking to the others.

Newland suddenly realized that he had said nothing to Ellen since the dinner began. He had to speak to her, to have a polite conversation.

1. **deserted** : left; abandoned.
2. **slender** : thin and graceful.
3. **scenes** : (here) public displays of anger or emotion.
4. **merciless** : very cruel.

"How was your journey from Washington?" he asked.

She looked at him, and her eyes said clearly, "Oh yes! Let's play our parts well!" "It was fine," she replied, "but the train was very hot."

"You won't have that problem in France," he said. "I remember one train ride from Calais to Paris. I've never been so cold in my life!"

She laughed.

"But no matter how uncomfortable it is, travel has its advantages," said Newland, raising his voice to address Larry Lefferts on the other side of Ellen. "You get away — you see something new. I'm planning to do some traveling myself soon. Hey, Larry, let's take a trip around the world, starting next month!"

"I can't go next month," replied Lefferts. "There's the charity ball for the blind. I can't miss that."

At this point, the ladies went to the drawing room. The gentlemen lit their cigars and returned to their conversation about Beaufort. "All the old values are changing now," said Lefferts. "In a few years we'll all be marrying our children to Beaufort's bastards!"

"Oh, dear!" cried Sillerton Jackson. "What a terrible thing to say!"

Henry van der Luyden sat at the end of the table, with an expression of sadness and disgust upon his face.

*

Two hours later, everyone left. Suddenly, Newland was by the front door and Ellen was in front of him, offering him her hand. "Goodbye," he said. "I'll see you in Paris." He spoke in a loud voice: he wanted everyone to hear.

"Oh," she replied. "I would be so happy if you and May could visit me there!"

Then she was gone.

He walked up the stairs slowly and went into the library. He lit a cigarette and stood gazing into the fire.

May came in and sat down in the big armchair beside him. "Well! I think that was a great success, don't you?" she said brightly. "Do you mind if I stay here with you and talk about the party?"

"All right." He sat down in the armchair opposite her. "There's something I'd like to discuss with you first. You see, I'm tired, May. Very, very tired."

"Oh, my dear! I thought so. You've been working too hard!"

"Perhaps. Anyway, I need a break —"

"You mean from the law? You want to give up the law?"

"I want to go away — to travel for a while — to get away from everything."

"Where to?" asked May.

"Oh, I don't know. India — or Japan."

"But I'm afraid you can't do that, my dear," said May gently. "Not unless you take me with you, and I don't think the doctor would let me go..."

She came over and put her arms around him. She was blushing and there were tears in her eyes. "I found out this morning that I'm expecting a baby!"

"Oh, my dear!" said Newland. He stood up and embraced her. She was warm and trembling in his arms. "Have you told anyone else?"

"Only Mamma and your mother," she replied. He couldn't see her face: it was buried in his shoulder. "Oh, and Ellen. I told Ellen. You remember I told you that we had a long talk and she was very kind to me."

"Yes. I remember. But wasn't that two weeks ago? I thought you said you just found out this morning..."

"Yes," said May. "It's true. I wasn't sure about it then, but I told her I was. And, you see, I was right!" She raised her head and looked at him, her blue eyes wet with victory.

The text and **beyond**

FCE 1 Comprehension check

For questions 1-6, choose the correct answer — A, B, C or D.

1 Newland asked May to go home with him because he
- **A** ☐ had a bad headache and needed to rest.
- **B** ☐ wasn't enjoying the play.
- **C** ☐ wanted to confess his love for Ellen.
- **D** ☐ no longer liked the company of Larry Lefferts.

2 When May said "it's all over" (page 107), "it" referred to
- **A** ☐ Ellen's problems with her husband.
- **B** ☐ Ellen's family's difficult situation with Ellen.
- **C** ☐ Ellen's allowance from her grandmother.
- **D** ☐ Ellen's friendship with Regina Beaufort.

3 Ellen sat next to Newland at the party because
- **A** ☐ she was his lover and everybody knew it.
- **B** ☐ he was the host and the party was for her.
- **C** ☐ he had helped her professionally.
- **D** ☐ he wanted to say goodbye to her.

4 New York society came to the party because they wanted to
- **A** ☐ tell Ellen how sorry they were that she was leaving.
- **B** ☐ show May that they liked the way she had saved her marriage.
- **C** ☐ make sure Ellen did not decide to stay in New York.
- **D** ☐ show Newland that they liked what he had done for Ellen.

5 Mrs van der Luyden mentioned Regina because she wanted
- **A** ☐ Newland to avoid being friendly with her.
- **B** ☐ Ellen to know that she made a mistake by talking with her.
- **C** ☐ Newland to know that May too will be punished if he continues seeing Ellen.
- **D** ☐ Newland to understand that Regina is a good wife just like May.

6 May had won her victory by
- **A** ☐ becoming pregnant.
- **B** ☐ lying to Newland.
- **C** ☐ lying to Ellen.
- **D** ☐ having a party for Ellen.

2 The custom of the city

For questions 1-10 read the text below and decide which answer — A, B, C or D — best fits each space. There is an example at the beginning (0).

The customs of that period were simple. The main (**0**) ..D.... activities of my father's friends were sea fishing, boat (**1**) .... and hunting. There were (**2**) .... clubs in New York in that period, and my mother, who was always very practical, said that people married early because the young men "had nowhere (**3**) .... to go". The young married (**4**) .... entertained one another often at their homes.
My readers, by this time, (**5**) .... be wondering what were the particular merits, private or public, of these friendly people. Their lives certainly do not seem to be very exciting. But I (**6**) .... that these people did something very important: they maintained the most complete honesty (**7**) .... business.
New York has always been a commercial community, and in my childhood the merits and defects of its citizens were (**8**) .... of a mercantile middle class. The first duty of such a class was to maintain a high level of honesty in business. I remember very well the horror excited by any dishonesty or irregularity in business affairs. I also remember how the families of those dishonest men were ostracized from society. Anyhow, I would (**9**) .... that the qualities that justified the existence of our old New York society were pleasant social (**10**) .... and total financial honesty, and in our modern society, which has gone so far away from these two qualities, we can easily see how much they are worth.

0	A fun	B relax	C free	D leisure
1	A running	B speeding	C competing	D racing
2	A none	B nothing	C any	D no
3	A else	B another	C other	D also
4	A partners	B pairs	C couples	D matches
5	A can	B could	C may	D have
6	A consider	B believe	C regard	D reflect
7	A in	B at	C of	D about
8	A those	B these	C they	D them
9	A tell	B inform	C speak	D say
10	A actions	B behavior	C manners	D performance

T: GRADE 8

3 Speaking: personal values and ideals

Edith Wharton wrote (in activity 2) how the New York society of her childhood valued highly good manners and personal honesty. Give a short presentation to your class about the values of your society. Use the questions below to help you.

1 What does your society think of politeness and good manners?

2 What about honesty in business?

3 Which of the following values and qualities are most appreciated in your society in general?
- honesty
- helping others
- generosity
- aggressiveness
- humility
- cleverness
- determination

4 What qualities are most appreciated by your friends?

4 Speaking – the 'hot seat'

A chair in front of the class is the 'hot seat'. While you are sitting in this chair you are either Newland, Ellen or May. The people pretending to be Newland, Ellen or May should try to be as accurate as possible, but they will also have to invent things and interpret things themselves. Take turns sitting there. The rest of the class can ask the person in the hot seat any questions. Below are some examples.

Questions for Ellen

- Did the hypocrisy of your family bother you at the party?
- Are you happy to be going to Paris?
- What kind of life do you want there?

Questions for Newland

- What do you dislike the most about New York society?
- How did you feel when you discovered that May had tricked Ellen?
- Will you ever go and see Ellen in Paris?

Questions for May

- When did you first realize that your husband was in love with Ellen?
- How do you feel about your victory?
- Do you think you and Newland will now have a happy life together?

A scene from Martin Scorsese's 2002 film ***Gangs of New York***. Daniel Day Lewis (centre, in red) is "Bill the Butcher", leader of the "Native Americans" (here it means people who were born in the USA, as opposed to first generation immigrants). On his right is Leonardo DiCaprio as the leader of the Irish immigrants.

The New York Films *of Martin Scorsese*

A gallery of portraits of Manhattan life

Martin Scorsese is one of the leading American film directors of his generation, and finally, in 2006, he received the Oscar for best director for his film *The Departed*. He was born in New York in 1942 to a family of Sicilian descent. He considered entering the priesthood, but then decided to go into film-making. He studied film at New York University and later taught there. Among his students were Oliver Stone and Spike Lee. When asked about his priorities in life, Scorsese replied, "My whole life has been movies and religion. That's it. Nothing else'."

Many Scorsese films are set in New York and characterized by relentless [1] violence. In fact, one can view his New York films as a gallery of portraits of the city, focusing on various periods and segments of New York society. *Mean Streets* (1973) is set in New York's Little Italy district. It stars Harvey Keitel and Robert De Niro, both of whom acted in several later Scorsese films. The sound track combines opera music with songs by the Rolling Stones – another typical Scorsese touch. In *Taxi Driver* (1976), New York City is one of the main characters, as we follow the strange life of a psychopathic loner. *New York, New York* (1977), with Liza Minelli and Robert De Niro, is about two musicians in Manhattan after the Second World War. *Raging Bull,* set in the Bronx in the 1940s and '50s, is based on the life of middle-weight champion boxer Jake LaMotta. The violence that brings him victory in the boxing ring also ruins his life. *After Hours* (1985) is set in Manhattan's Soho district and involves a string of encounters with bizarre New York characters. *New York Stories* (1989) is a feature-length film made up of three short films by different directors: Martin Scorsese, Woody Allen, and Francis Ford Coppola. Scorsese's contribution to New York Stories is called 'Life Lessons' and is about a painter in Manhattan. Its sound track consists of hit songs by Bob Dylan, Ray Charles, Procol Harum, and others. *Goodfellas* (1990) opens with the famous line, "As far back as I can remember, I've always wanted to be a gangster". The film is based on the lives of actual gangsters in New York's Italian mafia [2] in the 1960s and '70s. *Bringing Out the Dead* (1999), with Nicholas Cage and John Goodman, is about an ambulance driver in Manhattan. Its central themes are typical of Scorsese: New York as a city where

1. **relentless** : sth bad that never seems to stop or improve.
2. **mafia** : a secret organization of criminals.

Martin Scorsese's 1993 film ***The Age of Innocence***, with Daniel Day Lewis and Winona Ryder, offers an accurate portrait of the luxurious world of New York's upper class in the 1870s.

violence is part of the culture and conflicts are caused by ethnic[1] differences. *Gangs of New York* (2002), with Leonardo DiCaprio, Daniel Day-Lewis, and Cameron Diaz, is set in 1863. The film, based on historical material, explores corruption, violence, and gang warfare in New York as waves of immigrants from different countries of origin fight for positions in their new homeland.

The Age of Innocence (1993), with Michelle Pfeiffer as Ellen Olenska and Daniel Day-Lewis as Newland Archer, seems, at first, to be very different from Scorsese's other New York films. It focuses on the city's most privileged class at the end of the nineteenth century. The sensuous beauty of the film – the enjoyment of physical pleasures like fine cuisine, beautiful clothes, dancing, music, and elegant

1. **ethnic** : relating to different racial or cultural groups.

interior design, is heightened by the soundtrack, which is dominated by classical pieces by Strauss and Beethoven. But in this exquisite setting, social conventions cause intense pain to individuals who cannot escape their restrictions. Scorsese said that it was his most violent film, but the violence – this time – was psychological rather than physical.

1 Comprehension check

Say whether the following statements are true (T) or false (F), and then correct the false ones.

		T	F
1	Scorsese won an Oscar for his film version of *The Age of Innocence.*	☐	☐
2	He was born in Sicily.	☐	☐
3	He does not consider religion as an important part of his life.	☐	☐
4	Scorsese's *The Age of Innocence* is one of his several films about New York.	☐	☐
5	Scorsese's films are often very violent.	☐	☐
6	*The Age of Innocence* is Scorsese's only historical film.	☐	☐
7	In *The Age of Innocence* Scorsese gives a detailed picture of the luxurious life of the rich.	☐	☐
8	Scorsese sees *The Age of Innocence* as a calm and peaceful film of gentle beauty.	☐	☐

2 Speaking

If you have seen a film by Scorsese, talk about what you thought of it.

Before you read

1 Listening

FCE

You will hear about Newland nearly 26 years later; he is sitting in his library thinking about his past and present. For questions 1-6, choose the correct answer — A, B or C.

1 Newland's library had been the setting for the central events of his
- A ☐ business life.
- B ☐ artistic life.
- C ☐ family life.

2 His son, Dallas, was christened in the library because
- A ☐ he was not strong enough to be taken outside.
- B ☐ it was the tradition of old New York families.
- C ☐ the Bishop of New York did not want to do it in church.

3 His daughter's husband was not very
- A ☐ interesting.
- B ☐ dependable.
- C ☐ wealthy.

4 Newland's daughter adored
- A ☐ art.
- B ☐ sports.
- C ☐ traveling.

5 Dallas had a job with
- A ☐ a lawyer.
- B ☐ an artist.
- C ☐ an architect.

6 The fact that Dallas was going to marry Fanny Ring's daughter
- A ☐ shocked many.
- B ☐ bothered nobody.
- C ☐ interested only older people.

CHAPTER **ELEVEN**

Paris

Almost twenty-six years later, Newland Archer stood in his library, looking at the fire. His hair was now gray. This room had seen the most important moments of his family life. Here his wife had told him that she was expecting their first child — their eldest son, Dallas. Here Dallas had been christened by their old friend the Bishop of New York because he was too delicate to be taken to church. Here their second child, Mary — who looked very like her mother but was not as pretty — had announced her engagement to a boring and reliable young man from an old New York family. Here he and May had always discussed their children's futures. Mary had a passion for sports. Dallas was "artistic" and had finally found work in the office of an important New York architect. For the past six months, Dallas had been engaged to Fanny Beaufort. Lefferts had been right. Fanny was the illegitimate [1] daughter of Julius Beaufort and Fanny Ring, but no one seemed to mind about that now. She was a delightful young woman, and Newland was glad to welcome her into the family. Perhaps May might not have approved, but she had died two years before, so her approval wasn't necessary.

END

1. **illegitimate** : born to parents who are not married to each other.

But above all, it was in that library that his friend Theodore Roosevelt [1], who was then Governor of New York, had said to Newland, "You're the kind of man this country needs, if we're ever going to solve its problems!" Newland had tried to be elected to public office but without success. However, he continued his useful work for the city of New York and its people. For many years, for every public, artistic or philanthropic [2] project, people always wanted Newland's opinion. This was a big change for a man of his generation. Newland had escaped from the narrow world of old New York when men thought about nothing except sport, moneymaking and society.

Even so, Newland knew he had missed something: the flower of life. When he thought of Ellen Olenska, it was abstractly, serenely, as one might think of an imaginary beloved in a book or a picture. She had become the image of all he had missed in life, and that image had kept him from other women. He had been a faithful husband, and when May had died he had honestly mourned her [3]. Their long years together had taught him that it doesn't really matter if marriage is a dull duty, as long as it kept the dignity of a duty.

The phone rang.

"Hello, Dad?" said a lively young voice from Chicago. "It's Dallas. Listen: I'm leaving for Paris on Wednesday, and I want you to come with me. I have to go for business, but, if you come too, we can make it a holiday — our last father-and-son holiday before I get married."

Newland felt a little nervous. May hadn't liked traveling. Now he was used to that quiet life, and the idea of going abroad was a little frightening. But Dallas was right: it was their last chance to take a holiday alone together. "Yes, all right," he said, laughing.

1. **Theodore Roosevelt** : (1858-1919), President of United States (1901-1909); Vice President (1901); Governor of New York (1899-1900).
2. **philanthropic** : helping the poor and the needy, especially by giving money.
3. **mourned her** : felt sad about her death.

It was strange to be in Paris. For the first few years after Ellen had left New York, Newland had often thought of Paris as the setting of her life. He had imagined the horse-chestnut trees on the boulevards flowering in the spring, the great river rolling under its splendid bridges, the life of art and study and pleasure, and now here it was! He was walking down those boulevards with his son, and his heart was beating fast. Looking at Dallas, he wondered if his son's heart beat like that in the presence of Fanny Beaufort. He thought probably not. Can your heart beat fast for something that is permitted? He remembered the calm way Dallas had announced his engagement, absolutely sure that no one would object. "The difference between his generation and mine," Newland thought, "is that they assume they will get everything they want, whereas we almost always assumed that we would not."

"Isn't this lovely, Dad?" said Dallas, putting his hand on his father's shoulder as they walked along. "We've had the whole day to ourselves, but now we must hurry: we're going to visit Countess Olenska at half past five — Fanny made me promise to visit her. She said the Countess was so kind to her when she was in Paris."

Newland stared at him in amazement. "You told her I was here?"

"Of course — why not? Tell me, what was she like?"

Newland blushed and was silent under his son's curious gaze [1].

"Come on, Dad! You and she were great friends, weren't you? They say she was really lovely."

"Lovely?" said Newland. "I don't know. She was different."

"Ah, yes! That's how I feel about Fanny."

"What do you mean?" asked Newland.

1. **gaze** : fixed look; continuous look.

"Oh, Dad! You're so old-fashioned! Don't be so prehistoric! Why can't we talk about it? Didn't you once feel about her just like I feel about Fanny? She was the woman you really loved. I know. Mother told me the day before she died. She said she knew we would be safe with you because once, when she'd asked you to, you'd given up the thing you most wanted."

Newland walked on in silence. Then he said, "She never asked me."

"No — I forgot. You never did ask each other anything, did you? Or tell each other anything. You just sat and watched each other and tried to guess what the other was thinking. You're not angry with me, are you, Dad?"

"No — no, of course not."

And it was true: he wasn't angry with Dallas for saying those things. It was a relief to know that someone had guessed and pitied him, and it was terribly moving to know that it had been his wife. And he wasn't angry that Dallas had arranged for them to go and visit Countess Olenska. She had never returned to her husband, and, when he had died some years earlier, she hadn't changed her way of life. There was nothing now to keep them apart. "After all, I'm only fifty-seven," he thought.

They walked to the quiet square where Countess Olenska lived. The early evening light was golden. For thirty years, her life — of which he knew so little — had been lived in this golden light. Her life must have been full of interests — art, conversation, people — which he could hardly understand.

"Here it is," said Dallas, stopping in front of a modern building. "She lives on the top floor. Come on, it's almost time."

"I think I'll sit here for a while," said Newland. He sat down on a bench under a flowering horse-chestnut tree.

"Why — are you ill?"

"No. I'm fine. But, please, go up without me."

"You mean you'll come up later?"

"I don't know," replied Newland slowly.

"If you don't, she won't understand."

"Go on. I'll follow you."

Dallas stared at him. "But what shall I say to her?"

"My dear boy, you always know what to say."

"All right. Shall I say that you're old-fashioned and prefer to walk up five flights of stairs because you don't like lifts?"

"Just say I'm old-fashioned. That'll be enough."

Looking perplexed, Dallas turned and walked into the building.

Newland sat on the bench and gazed up at the balcony and the windows on the top floor. He calculated the time it would take Dallas to go up in the lift, ring the door bell, be admitted, and walk into the drawing room. He wondered if it was true, as people said, that Dallas took after him [1].

He tried to imagine the people in the drawing room. Six was the hour for visiting. He was sure there would be more than one person there. Among them would be a pale lady with dark hair. She would hold out her hand to Dallas. He thought she would be sitting near the fire. There would be flowers on the table beside her. "It's more real to me here than if I went up," he thought. He sat on the bench for a long time, as the golden light faded and the evening came. He watched the lights come on in the room behind the windows on the top floor. Then a maid came out and closed the shutters.

Newland Archer got up from the bench and walked back alone to the hotel.

1. **took after him** : looked like he did when he was young.

The text and **beyond**

1 Comprehension check

Answer the questions below.

1 What were Newland's two children like?
2 How had Larry Leffert's prediction come true?
3 How had Newland spent his time?
4 What had Ellen come to represent for Newland?
5 Why did Newland hesitate about going to France?
6 How did Dallas convince him?
7 What does Newland think is the difference between him and his son?
8 What does Dallas think is the difference between him and his father?
9 How are Fanny Beaufort and Ellen similar according to Dallas?
10 What, according to May, was Ellen for her husband?

2 Writing: my old-fashioned friend

Pretend you are Ellen. Write to a friend about your visit from Dallas and about Newland's refusal to come and see you.
In your letter include:

- a brief description of Dallas
- how you felt when Dallas wrote to you that he was coming with his father
- how you felt when Dallas appeared without his father
- why you think Newland decided not to come up

3 Summary

Number the paragraphs in the right order to make a summary of chapters Seven to Eleven. Then fill in the gaps with the words in the box below. One has been done for you as an example.

get	**expecting**	**stroke**	**back**	**victory**	**lovers**	**carriage**
old-fashioned	**museum**	**squalid**	**bankrupt**	**far**	**lend**	
independent	**backs**	**trusted**	**near**	**headache**		

A ☐ All of the elite of New York was at the party, and as Newland looked at them he realized that they all believed that he and Ellen had been .... After the guests had left, they went up to the library. Newland began to tell May that he wanted to .... away from everything. May told him that this was impossible: she was .... a baby. This was May's moment of ....

B [1] Julius Beaufort went .... and it was a terrible scandal in New York. His wife Regina went to ask Mrs Manson Mingott to .... her husband some money. Mrs Mingott refused. However, the stress was too much for Mrs Mingott and she had a ....

C [ ] Nearly 26 years later Newland and his son Dallas were in Paris. That evening he and Dallas were supposed to visit Ellen. But Newland decided not to go up. He asked Dallas to tell Ellen that he had decided not to come up because he was ....

D [ ] The next evening May and Newland were at the New York Academy of Music. Newland told May that he wanted to return home because he had a ....

E [ ] Six or seven days later Mrs Mingott called for Newland. She told him that Ellen was going to stay with her, and that this was against the family's wishes. They wanted her to go .... to her husband.

F [ ] The next day Newland met Ellen at the .... — the only place in New York where they could be alone. Again they discussed their situation. As an act of desperation, they decided that they would be lovers once.

G [ ] Later Mrs Mingott felt better. She wrote to Ellen and told her to return to New York. Ellen was arriving at Jersey City, so Newland went to meet her in a ....

H [ ] During the ride to Mrs Mingott's house, they talked about their love. They agreed that they did not want a .... affair. Ellen told him that they were .... each other only if they stayed .... from each other. She also said that she did not want to be happy behind the .... of the people who .... her.

I [ ] When they were in the library Newland started to tell May about Ellen and him, but she interrupted. She said there was no reason to talk about Ellen because she was returning to Europe. Mrs Mingott was giving her enough money to be .... of her husband. She concluded by saying that they should have a farewell party for Ellen.

INTERNET PROJECT

The Progressive Era

Newland Archer, as we have seen, began to participate actively in the world around him. He was following the example of Theodore Roosevelt (who was a real-life friend of Edith Wharton). This period from 1900 to 1917, the year of the United State's entrance in World War I, is known as the Progressive Era. This was the period when many people in government tried to face some of the serious social problems affecting the nation.

To find out more about the Progressive Era, go to the Internet and go to www.blackcat-cideb.com or www.cideb.it. Insert the title or part of the title of the book into our search engine. Open the page to *The Age of Innocence*. Click on the Internet project link. Scroll down the page until you find the title of this book and click on the relevant link for this project.

Prepare a short report on one of the following aspects of the Progressive Era:

- big companies and trusts
- workers' rights
- women's rights
- life in the cities
- Theodore Roosevelt
- child labor
- immigrants

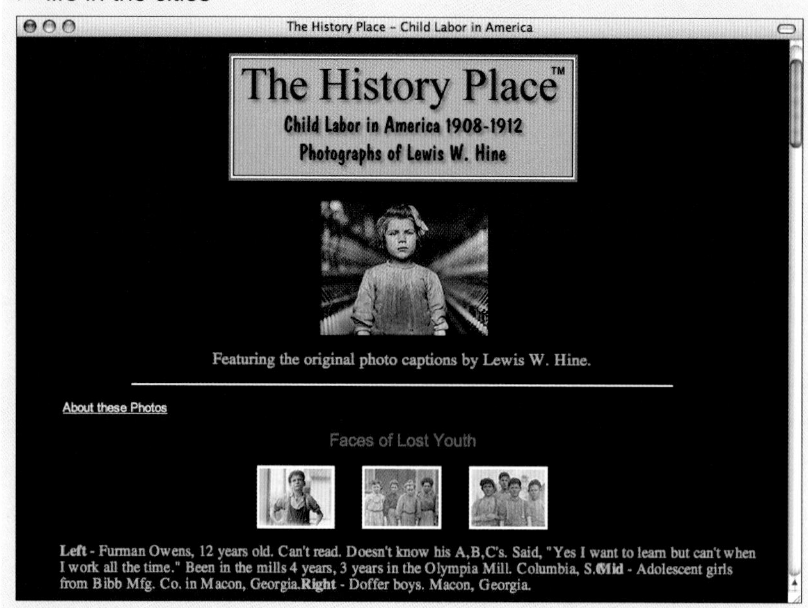

EXIT TEST 1

1 Characters

Who said what and why? Match the quotes with the characters who said them, and then match the quotes with the reason why they said them. You may match some characters with more than one quote.

Who

Newland (N)	May (M)	Mrs Mingott (Mg)
Mr Letterblair (L)	Sillerton Jackson (S)	Ellen (E)

What

A ☐ ☐ "I hear he was still helping her a year later."
B ☐ ☐ "Why wait?"
C ☐ ☐ "Do you want that family to be the subject of scandal?"
D ☐ ☐ "Julius's business keeps him in the city most of the time."
E ☐ ☐ "I've arranged to go to a farm in the north to look at some horses."
F ☐ ☐ "The darling!"
G ☐ ☐ "Catch my death!"
H ☐ ☐ "I won't open my mouth unless you tell me to."
I ☐ ☐ "So I could ask you, my dear boy, what do you mean by asking me what I mean?"
J ☐ ☐ "What a coincidence!"
K ☐ ☐ "But it doesn't matter now, does it, now that it's all over?"

Why

1. He is telling Newland why he should convince Ellen not to get divorced.
2. She is announcing her victory over her husband's lover.
3. He is implying unpleasant things about Ellen.
4. He is making an excuse so he can see where Ellen is staying.
5. She is making an ironic allusion to the fact that Beaufort has a mistress.
6. He is feeling his first doubts about marrying May.
7. He is asking Ellen passionately to have lunch with him.
8. She is letting Newland know that she knows he was lying in order to be with Ellen.
9. He is happily impressed by May's innocence.
10. He feels angry about being shut up in a stifling marriage.
11. He is saying that he knows that Newland is jealous of Beaufort.

2 Picture summary

Look at the pictures from *The Age of Innocence* below. They are not in the right order. Put them in the order in which they appear in the story.

3 A graphic novel

Photocopy these two pages, cut out the pictures and stick them on paper in the right order. Think of words to put in speech or thought bubbles to show what the characters are saying or thinking. Do not use the words that were used in this book! Then write at least one sentence under each picture to narrate what is happening.

EXIT TEST 2

1 Answer the following questions.

1 What did Newland think of May's innocence before they were engaged to be married?
2 How did Newland see May's innocence after he had fallen in love with Ellen?
3 What impression did Ellen give Newland when they first met at the opera?
4 How did New York society show its disapproval for the public appearance of Ellen Olenska?
5 How did Newland persuade Ellen not to divorce her husband?
6 How did May interpret Newland's request for an early wedding?
7 How did Newland see Ellen during their honeymoon?
8 How did Newland see Ellen once they had returned to New York from their honeymoon?
9 What caused Mrs Mingott's stroke?
10 What was the real disgrace of Julius Beaufort according to Mrs Mingott?
11 How did New York society see the farewell dinner for Ellen held by Newland and May?
12 How did May see Newland's giving up of Ellen?
13 What was Larry Lefferts' prediction about the future of New York high-society families?
14 What excuse did Newland give for not going up to see Ellen in Paris?

2 **Say whether the following questions are true (T) or false (F), then correct the false ones.**

		T	F
1	At the time of our story, New York high society was a fairly small group of people.	☐	☐
2	Newland and May announced their engagement earlier than planned to protect Ellen.	☐	☐
3	Ellen immediately understood that New York society did not approve of her.	☐	☐
4	Newland was surprised at how conventional Ellen's house was.	☐	☐
5	Julius Beaufort and Ellen both had European points of view.	☐	☐
6	Ellen thought she could divorce her husband because her family was Protestant.	☐	☐
7	Newport was where most of the men of New York high society worked.	☐	☐
8	Ellen went to Boston to talk with her husband's secretary about an offer her husband had made.	☐	☐
9	Ellen told Newland she would stay in America only if they became lovers.	☐	☐
10	In the end, Mrs Mingott agreed with her family that the best thing for Ellen was to return to her husband.	☐	☐
11	May did not realize that her husband and Ellen were having a kind of love affair.	☐	☐
12	May was not certain that she was pregnant when she told Ellen about it.	☐	☐
13	After Ellen left for Europe, Newland lived a very private and closed life.	☐	☐
14	Newland's son knew all about his father's great passion for Ellen Olenska.	☐	☐

A Brief History of Divorce

In ancient Mesopotamia, Athens, and the Roman Empire, divorce was accepted, although it happened rarely. With the advent of Christianity, however, divorce was restricted to very serious causes and, by the tenth century, it had almost disappeared, because of the influence of the Church, which considered marriage a sacrament that was made by God and could not therefore be dissolved by humans. After the tenth century, married couples sometimes separated, but they were not allowed to marry again and the husband was obliged to continue to support his wife. The only way a marriage could be dissolved (permitting remarriage and releasing the husband from the responsibility of supporting his wife) was by annulment, in which a priest declared that there were reasons, recognised by the Church, why the marriage should never have taken place.

Perhaps the most famous and important divorce of all time was that of King Henry VIII of England from his first wife Catherine of Aragon. In 1533, having failed to get the Pope's permission for a divorce, Henry married Anne Boleyn. The Archbishop of England declared that the King's marriage to Catherine was not valid but that his second marriage to Anne was. The Pope reacted by excommunicating Henry. In 1534, the English Parliament passed the Act of Supremacy, which declared that the King was 'the only Supreme Head in Earth of the Church of England'. Thus, because of one man's dissatisfaction with his wife, England separated from the Roman Catholic Church, and Protestantism, which was already gaining ground in Europe, became the dominant religion of England and its future settlements (Canada, America, Australia). Thereafter, opposition to divorce was seen as one of the major characteristics that divided Protestants from the Catholic Church, but in fact, though their Church allowed it, Protestants around the world continued to disapprove of divorce until well into the twentieth century.

At the end of the nineteenth century – when *The Age of Innocence* is set – divorce was legal but was disapproved of in polite American society. If you got divorced, your personal life became a scandal, and 'respectable' people would probably talk about you a lot but were not likely to talk to you or invite you to their dinner parties. Edith Wharton herself had suffered from this. Her marriage had never been a happy one (at the age of twenty-three she had married a rich banker of thirty-eight), but, when her husband became mentally ill, the marriage became unbearable. Edith reacted by dividing her time between her husband's home in Massachusetts and her house in Paris. In 1913, Edward Wharton was committed to an insane asylum, and Edith obtained a divorce and lived the rest of her life in Paris. The decision to live in Paris was partly a personal preference (Paris was far more intellectually and artistically stimulating than New York) and partly imposed upon her (as a divorcée in New York, her life would have been very difficult indeed).

These days, in the USA, between 40% and 50% of marriages end in divorce. This relatively high rate is thought to be the result of no-fault divorce (in which divorce can be obtained without citing 'grounds' such as adultery or abandonment, if both spouses agree). The kind of pain that Edith Wharton and Ellen Olenska experienced is therefore no longer as common as it was in the late nineteenth and early twentieth centuries.

The Age of Innocence — Key to the Exercises and Exit Tests

KEY TO THE EXERCISES

The Life of Edith Wharton

Page 9 – activity 1

1 False, they were fairly wealthy.
2 False, it was about the decoration of houses.
3 False, several of her books were quite popular.
4 True
5 True

CHAPTER ONE

Page 11 – activity 2

1 was too old and fat
2 daughter
3 understand
4 are getting married
5 love song
6 high society's scandals and secrets

Page 18 – activity 1

1 It is not very important to him.
2 She is May's cousin.
3 He thinks she's right to help her, but that she should not have brought her to the opera.
4 Lovell Mingott brought her there from Venice.
5 She had married a wealthy Polish count who treated her badly.
6 She ran away from her husband with her husband's secretary.
7 To show everybody that he supported May and her family's decision to help Ellen.
8 She wore bright red silk soon after her parents' death and not black.
9 Ellen's aunt.
10 He was a horrible man.
11 She remembers that he kissed her.
12 She spoke about New York flippantly in a sophisticated European way that annoyed Newland.

Page 19 – activity 3

1 done **2** it **3** own **4** any
5 got/found **6** like **7** kind **8** such
9 would **10** in **11** made/built **12** had

CHAPTER TWO

Page 20 – activity 1

1 shuttle **2** list **3** skull **4** sung
5 mean **6** speed **7** piece **8** term
9 rocks **10** beyond

Page 28 — activity 1

1 J **2** K **3** H **4** B **5** E **6** A **7** D
8 G **9** I **10** F

Page 29 — activity 2

1 gathering **2** founding **3** known
4 fashionable **5** privileged
6 certainly **7** decoration
8 unbelievable **9** organizer
10 snobbish **11** refined **12** American

Page 30 — activity 3

1 business - the others are names of jobs or professions.
2 New Yorker - it is the only one of the three that can only be used for people.
3 Gloves - the other three refer to jewellery.
4 see, which means 'to perceive with your eyes' - the other three mean that you decide to direct your eyes at something.
5 Hairstyle - the other three are all things that you actually wear.
6 Mistress - the other three all relate to the communications of events or scandals.
7 Relatives - the other three are all connected to the idea of people being or getting married.

CHAPTER THREE

Page 30 — activity 1

1 Artists and writers
2 Italy
3 Yes, he did.
4 Very artistic and it smelled of spices - it was very different from a usual New York drawing room.
5 Two
6 New Yorkers never had less than a dozen.
7 May and her mother

Page 36 — activity 1

1 A **2** B **3** B **4** B **5** D **6** A

Page 37 — activity 2

A 6 **B** 11 **C** 8 **D** 2 **E** 9 **F** 1 **G** 4 **H** 3
I 5 **J** 10 **K** 7

New York in the 1870s

Page 42 — activity 1

1 Until 1790.
2 In 1850 it was 696,000, and in 1900 4 million.
3 New Amsterdam.
4 Supposedly, the 400 most important people in the city.
5 He considered every person in the city important, not just the 400 wealthiest.
6 There was massive migration of African Americans from the south, the beginning of the construction of the skyscrapers and the digging of the subway system.
7 The calmer days before the dizzying growth of the early twentieth century.

CHAPTER FOUR

Page 42 — activity 1

1 Mrs Manson, Ellen's grandmother.
2 They want him to convince her not to seek a divorce.
3 It is a very private matter, and Newland is about to become part of the family.
4 Money is not an issue: the count has already given some money back, which he was not obliged to do, and Ellen does not seem to be very interested in money.
5 Create a scandal.

Page 49 — activity 1

A L 4 **B** L 6 **C** B 3 **D** N 9 **E** E 10
F E 1 **G** M 8 **H** N 5 **I** E 2 **J** E 7

Page 50 — activity 2

1 D **2** C **3** A **4** H **5** B **6** F

Page 52 — activity 3

CHAPTER FIVE

Page 62 — activity 1

1 F - He was actually afraid of doing something foolish since he was in love with Ellen.
2 T
3 F - She says that it was his fault - he had persuaded her not to divorce.
4 F - She says that she no longer feels lonely and so she has no need of his company.
5 P
6 F - He finally accepted their marriage once they were back in New York from their honeymoon.
7 F - He knew of her basic movements and actions.
8 F - He was beginning to lose money.
9 F - On the contrary, he was greatly affected by her presence.
10 F - She thinks Ellen should return to her husband.

Page 63 — activity 2

1 should have accepted
2 she shouldn't have spent the afternoon
3 she shouldn't have lived so long
4 she should have married
5 you shouldn't have advised
6 she should have showed more respect
7 she shouldn't have chosen

CHAPTER SIX

Page 73 — activity 1

1 C **2** C **3** B **4** C **5** A

Page 74 — activity 3

1 only thing I was
2 had turned down
3 would protect me
4 you were the kindest
5 she would not be
6 can only join
7 first time I have ever

8 have all been invited
9 had nothing to do
10 am the only one

Page 75 — activity 4

1 Newland would take care of the divorce for her
2 be married sooner
3 was in love with another woman.
4 she didn't want to hurt May and her family.
5 her parents had agreed that she and Newland could marry after Easter.
6 a dream from the past
7 had suffered greatly from being away from Newland

CHAPTER SEVEN

Page 76 — activity 1

1 Newland's mother's house.
2 dinner
3 he didn't spend all his money on Regina.
4 accepted her husband's offer.
5 calm
6 money
7 Julius Beaufort
8 they join the ladies.

Page 82 — activity 1

1 He was alluding to Julius Beaufort's mistress and all the money Beaufort had spent on her.
2 He believed - and he is partly right - that Mr Jackson was insinuating that Ellen and Julius Beaufort are having an affair.
3 He said that he was simply referring to the fact the Ellen had invested money with Julius.
4 He said he had business there.
5 Regina's asking her for a loan to save her husband, Julius Beaufort, from the dishonor of bankruptcy.
6 Because she and Regina are members of the same family.
7 She said no because Beaufort lost the money of people who trusted him. Also, Regina was a Beaufort when things went well and she is still a Beaufort now that things are going badly.
8 Ellen
9 She seemed to be quite suspicious of the sudden cancellation of his business appointment.

Page 82 — activity 2

1 increase **2** another **3** Between
4 even **5** owned **6** amount
7 manage **8** deals **9** occurred
10 involved **11** lasted **12** reached
13 blame **14** becoming **15** more
16 came

CHAPTER EIGHT

Page 85 — activity 2

1 C **2** B **3** A **4** C **5** C **6** A

Page 92 — activity 1

A N 6 **B** E 1 **C** M 8 **D** E 4 **E** N 7
F N 2 **G** E 5 **H** N 3

Page 93 – activity 2

MARRIAGE
ceremony engaged marry honeymoon

THE SEA
lighthouse sailboat pier

BEING SURPRISED
shocked alarmed startled

LOVE OUTSIDE MARRIAGE
affair mistress

SCANDAL
disgrace shame dishonor divorced

TRANSPORTATION
carriage station railway

ILLNESS
stroke cold

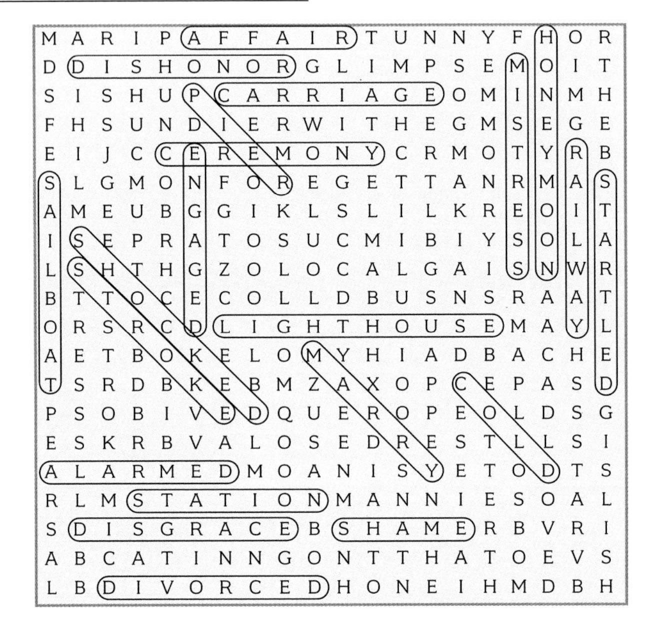

CHAPTER NINE

Page 94 – activity 1

1 F **2** F **3** T **4** T **5** F **6** T

Page 103 – activity 1

A 1 **B** 13 **C** 7 **D** 2 **E** 5 **F** 9 **G** 4 **H** 12 **I** 11 **J** 3

Page 104 – activity 2

1 Seventeen **2** Fifth Avenue **3** the lady in the carriage **4** told her not to stare at strange people **5** the first fashionable mistress in New York **6** looked out the other window **7** a mirage of palm trees

Recording script

I remember when a very small bright yellow carriage with a very elegant driver suddenly appeared in Fifth Avenue. I was 17 at the time. Inside this carriage I could barely see a lady. I have a vague memory of her. She had dark hair and wonderfully pale skin. It was the most elegant vision that had ever appeared on Fifth Avenue. But